I0488779

Over Our Heads

AN EXPLORATION INTO LIFE, THE UNIVERSE AND EVERYTHING

James Sinclair

Tmesis Publishing

Over Our Heads

A Exploration into Life, the Universe & Everything

Second Edition 2015

Front cover image : Vintage style Solar System ©Mike McDonald via Shutterstock

Book Layout ©2013 BookDesignTemplates.com

Cover design, formatting and general creative steerage by Tom Evans, aka The Bookwright

www.tomevans.co

TMESIS PUBLISHING
SURREY, UK
WWW.TMESIS.CO

ISBN: 978-1519773050

tmesis
AUTHOR-CENTRIC PUBLISHING

Foreword

To understand the very big, we have to understand the very small. To know the point of the very small, we have to see it in the context of the very big.

Compared to an atom, each of us is enormous. Compared to a human being, the Earth itself is gigantic.

Yet, our Earth is only a mere pale blue dot when viewed from the edge of our solar system. Try and observe our solar system from our nearest galactic neighbour and you would be looking for a needle in a haystack.

Perhaps the biggest mystery of all though is why we are conscious and self-aware enough to hold these thoughts in the first place.

James Sinclair has taken on the most daunting of tasks. As a passionate and natural teacher, he has written a relatively Small Book that explores Big Ideas. Over Our Heads delves into the Very Small before taking a look at the Very Big. Both of these realms are over the heads of most people. So, most importantly, he explains huge concepts in words readers without PhD's can understand.

While most people on this special planet go about their days not looking up, James reveals that the mysteries over our heads and beneath our feet are where we must look to find our true nature and the answer to why we are here.

Tom Evans : Author of This We Know and This We Are

Contents

[1]

A Brief Introduction

HOW OFTEN HAVE YOU SAT THROUGH SOME SCIENTIFIC DOCUMENTARY AND BY THE END OF IT BEEN QUITE "MIND-BOGGLED" AND CONFUSED?

All the talk about black holes, worm holes, warping of space and the Big Bang, might not mean very much to the average viewer. They might end up picking up the remote and changing the channel to something more comprehensible and less mind-numbing.

It is my hope and aim is that this little book will help the reader to make better sense of it all. Over Our Heads is aimed at those who have an interest in science but would like to know a bit more. I have attempted to present this in a way which would be more appealing to the 'unscientifically-minded.' For example, to explain very large distances so they will be more understandable to the average person, by comparing the relative spacings between planets in the Solar System to a scale we can imagine.

I have tried to steer away, wherever possible, from scientific names and tried to explain things in a simple, light-hearted manner.

Writing about the Universe is a very challenging project, because it means that you are trying to talk about everything that is contained within it, from the very smallest – an atom, to the very largest – a galaxy. Here one has to use one's mind, because you can neither see an atom nor reach out and touch a galaxy. We have to try and comprehend them only with our minds, and this is stretched to its very limit when we try to understand the vastness of space and the infinity of time.

The great mystery of life is also very difficult to explain. How we came to be, made up from basically the simplest of atoms that were forged in stars to become other chemical elements, to evolve into the complicated molecules of life and eventually into us – the children of stardust.

We are virtually the end-product of the Universe, the most complicated thing it has ever produced. Why we are here and how we came to be is probably the most thought provoking question we can ever ask ourselves.

Life, in its many shapes and forms on this unique planet, is, perhaps, the greatest rarity in this vast Universe. Are we the only ones? Can this minuscule pale blue dot when viewed from the fringes of the Solar System be the only habitable place for life to evolve?

These are very BIG questions for a little book like this to answer. But think of them we must and that is what it is intended to do.

I hold no degree in Science, but am just a mere mortal, obtaining my understanding of the Universe by reading books and articles on science, and from watching those "over our heads" science documentaries on TV that you may have so far avoided. But if I can give my reader a better understanding of the vast Universe and all its wonders, I will rest my case.

[2]

Tiny, Tiny

THE UNIVERSE IS ALL ABOUT SIZE, FROM MINUSCULE ATOMS, WHICH MAKE UP EVERYTHING AROUND US AND OVER OUR HEADS, TO SUPER-GIANT STARS THAT LIVE IN GREAT GALAXIES WHICH ARE THE LARGEST THINGS IN THE VISIBLE UNIVERSE.

Our starting point, therefore, begins appropriately with the very, very tiny – the stuff everything is made from.

THE ATOM

The concept of the atom is not something new; in fact the name comes from the ancient Greek 'atomos' - the ultimate building blocks of nature. It was conceived by early Greek philosophers like Democritus (460 – 370 B.C.) who was the first to coin the term atom which means indivisible in Greek, although their concept was rather different to what we now understand.

Everything in the Universe is made up of atoms - yourself, this book you're reading, the chair you are sitting on and the neighbour's cat.

Atoms are tiny, tiny particles which are more imagined than seen. But exist they do - consider the explosion of a nuclear bomb! Atoms are made up essentially of the nucleus, and much tinier particles called electrons, frantically orbiting around in a sort of "cloud" or "shell." The nucleus of the atom is made up of protons, and sometimes of protons and neutrons combined. Protons have a positive electrical charge while the neutron has no charge at all. They are held together in a tight cluster by the strongest force in nature, called the "strong nuclear force."

The other part of the atom, the electron, which circles round the nucleus has a negative charge, and is much, much smaller than the nucleus. The relative size of the nucleus to the electron cloud is that of a grain of sand (the nucleus) in the middle of the Albert Hall (the shell of electrons); so most of the atom is really empty space (like our Solar System). Atoms come in all forms - the simplest, is the hydrogen atom with just one proton and one electron circling round it, to the more complex uranium atom, which has 92 protons in the nucleus and 92 electrons orbiting. This balance of positively charged protons and negatively charged electrons cancel each other out to give the atom a neutral electrical status.

Hydrogen is the most abundant element in the Universe, making up more than 92 percent of the mass, for it is the matter that the stars we see are made of. The next most abundant element is helium, (over 7 percent) which, like hydrogen, is very light. Other elements like oxygen, nitrogen and carbon, which are essential to life, make up the rest of the matter in the Universe.

Atoms themselves are made up of smaller sub-atomic particles, and it is to determine what these particles are that particle accelerators, like the Large Hadron Collider (LHC for short) have been constructed. The LHC is an enormous 17 mile long circular tunnel sunk beneath the Franco-Swiss border where atoms are held in a strong magnetic field and beamed towards each other at near light speed from opposing directions. When these atoms collide, other sub-atomic particles from which they are composed, are ejected, and can be detected with specialised equipment. One of these particles is the mysterious Higgs boson. Theorised to exist as far back as 1964, it was only on March 14, 2013 that the Higgs is tentatively confirmed to exist. The Higgs boson is sometimes referred to as the "God Particle", because it is thought it is the Higgs that gives some fundamental particles their mass. However, it is a very elusive particle and decays almost instantly. It is named after Peter Higgs, one of six physicists who first suggested the existence of such a particle, and is considered so important that it was one of the primary reasons for the construction of the LHC.

THE ELEMENTS

No, we're not talking about the weather. We're talking about atoms. When atoms of the same type bond together they form an element.

An element is a substance which can't be broken up into any more simple particles of matter (except perhaps by nuclear reactions!) and they 'sort-of' exist in their own right. There are only 92 elements that occur naturally in nature and each has a chemical name and an atomic number.

The atomic number is determined by the number of protons in the nucleus. Some of the elements which we are more familiar with are sodium, calcium, iron, carbon, copper, silver and gold. In ancient times it was always the alchemist's dream that he could change base metals into gold. Unfortunately, he became unstuck and frustrated in his efforts, because gold is an element, and can't be made from something else! Having said that, however, one element can change into another under very high temperatures of millions and millions of degrees. This process is called nuclear fusion and is the same reaction as is found in the heart of stars or the explosion of a hydrogen bomb.

FROM GAS TO GOLD

Most of the Universe we see around us is made up of gases mainly hydrogen and helium. So you may well ask; "How and where do all the other familiar elements we see around us come from?" The simple answer is that they are forged in the anvils of stars. When a main sequence star, like our Sun, has converted all the hydrogen in its mass, to helium, releasing vast amounts of energy in the process, it has no other recourse but to then begin converting the helium to the next heaviest elements, carbon and oxygen. At this stage, the star begins to distend into a red giant.

When this happens to our star, it will bloat out till it swallows up the inner planets, our Earth included. Finally the star's atmosphere will float out into space in a ring, creating a pretty planetary nebula, with the core (a white dwarf) left shining in the centre. But more massive stars than our sun end their lives more spectacularly.

Their mass allows elements heavier than carbon and oxygen to be converted to further heavy elements, but when it comes to iron, this very stable element will not fuse into any other. Hence, nuclear reactions in the core cease, and since there is now nothing to hold the star up, it collapses under its own enormous gravity, in a matter of seconds, with a cataclysmic explosion called a supernova.

Under the enormous temperatures and pressures of this event, heavier, rare and precious elements like gold, silver, platinum and uranium are forged, so the gas – in the shock-wave of the explosion – is enriched with atoms of these heavier elements that we are familiar with on Earth.

Eventually this residue of the exploding star collides and combines with other nebulae, and from these "star nurseries" – like the Eagle and Orion nebulae – condenses into other stars. All that is left of this supernova explosion is the core of the massive star, a small, dense and extremely massive object known as a neutron star or possibly a pulsar. Supernovae have been observed in other galaxies, and the nearest one since the Crab was in the Large Magellanic Cloud, a satellite galaxy of our own Milky Way in 1987, called Supernova SN1987A,

Some unstable red supergiant stars, like the familiar Betelgeuse in Orion, 600 light years away, might explode as supernova at any time. However, if this event were to happen today, people on Earth would not witness it until the 27th century A.D. as this is the amount of time it would take for the evidence of this event to reach us, and it would light up our night sky in a blaze of light.

So, different types of stars (according to their mass) die in different ways. As in the case of our sun, imagine releasing the air out gently from a balloon, but in the case of supermassive stars, imagine pricking the balloon with a pin!

In conclusion, all the familiar objects we see around us, including each other, are made up of elements forged in the death throes of stars. So we are simply children of the stars and composed of stardust. This applies even to the iron in our blood (we have about enough to make a small nail). The same goes for all the life around us – both animal and vegetable – made mainly from carbon, as is the humble lead in your pencil.

MOLECULES AND CHEMICAL COMPOUNDS

Atoms are quite sociable and readily combine with other types of atoms to form molecules of compounds. For example, when the atoms of iron, combine with the atoms of oxygen, a compound – rust (iron oxide) - forms. This is known as a chemical change, where the properties of the compound (rust) are quite different from the elements iron and oxygen from which it is made.

Another example of a chemical change is the highly reactive metallic element, sodium, and the poisonous gas, chlorine, which form a very familiar compound, common table salt.

Water (which can exist as a liquid, solid, and gas) is probably the most familiar and abundant compound on Earth. On the other hand, substances can mix non-chemically too. For example a mixture of iron filings and sand are not a chemical change, because you can easily see the iron filings in the sand and separate them again with the help of a magnet. These mixtures are called a physical change, and air, a mixture of mainly nitrogen, and some oxygen, and other gases, is a prime example of a mixture.

Mixtures and compounds make up most of the familiar things around us such as the air that we breathe and the water we drink. Chemists have a unique notation system designed to simplify the writing of chemical names, such as water (H2O - meaning 2 atoms of hydrogen and 1 atom of oxygen), carbon dioxide (CO2 - 1 atom of carbon and 2 atoms of oxygen), and salt (NaCI - 1 atom of sodium and 1 atom of chlorine).

The most common, abundant and simple element is hydrogen – just a single proton (the nucleus) with a single electron buzzing round it.

In fact, almost everything we see in the Universe is made of hydrogen – stars, galaxies, and most of the matter in nebulae. In fact 75% of the entire Universe is composed of hydrogen. Paradoxically, though hydrogen is the most abundant element in the Universe, it is rarely found on Earth in its natural state and has to be made in a laboratory. Many schoolchildren would have participated in this experiment.

The next most abundant element in the Universe is helium, but unlike hydrogen, which was created as the first element in the Big Bang, helium is manufactured in stars. When enough hydrogen is accumulated, gravity, and the enormous heat and pressures it creates, forces the hydrogen gas cloud to ignite, and convert hydrogen into helium.

This process is called nuclear fusion and is what happens in a hydrogen bomb blast.

I hope I have not bored the reader with all this talk of the elements. If I have, I apologise and perhaps should move on to something more tangible like the weather!

[3]

Forces to be Reckoned With

THE THREE FUNDAMENTAL FORCES IN NATURE ARE THE STRONG NUCLEAR FORCE, THE ELECTRO-MAGNETIC FORCE AND GRAVITY.

Of these, the first is the strongest and the last, gravity, the weakest. To understand this, place a paper-clip on a table. The gravity of the entire Earth is holding the paper-clip to the table. Now take a small magnet off your fridge, and hold it an inch away from the paper-clip. The clip will fly up and attach itself to the magnet. So, you can conclude that the gravitational force of an object as large as the Earth can be overcome by the electro-magnetic force of something as small as a magnet held in your hand. However, weak as gravity may be, it is a force to be reckoned with when it comes to very massive objects in the Universe – supergiant stars, neutron stars and black holes. In fact it is gravity which apparently holds the whole Universe together.

On the other hand, the strong nuclear force is the force that holds atoms together, and is very powerful. It requires a large amount of energy to split the atoms of uranium (a very unstable and radioactive element) in a process called nuclear fission, which uses atoms to smash into other atoms in a chain reaction – much like a snooker player smashes into a pack of billiard balls and splits them with the cue ball. The cue ball in this instance would represent a neutron which smashes into the nucleus of an atom and knocks out other neutrons, which in turn knock into other atoms, splitting the atoms as they go. It is this chain reaction which causes the explosion of an atomic bomb – atoms smashing into atoms, and splitting them, in an uncontrolled way. Imagine you have a large container of about 100 mouse-traps, all sprung and ready to go. If you were to trigger one of the mouse-traps, it would cause the neighbouring mouse-traps to snap, and those in turn would trigger off their neighbouring mouse-traps till, in a few seconds, all the mouse-traps would be sprung in a runaway fashion.

Nuclear fission is in use today in nuclear reactors of power stations, used to heat water, to produce steam, to turn turbines, to produce electricity. The nuclear rods (which contain uranium and are radioactively "hot"), are submersed in a container of water and their energy released slowly in a controlled way to heat it. This is more energy efficient than coal-fired furnaces to do the job, but is more hazardous due to the radiation emitted and the nuclear waste, that has to be disposed of carefully.

Nuclear power plants can be dangerous too (remember Chernobyl or the more recent Fukushima incident after the earthquake in Japan), because if the core (the rods) is not kept cool by a constant supply of water or gas, it becomes hotter and hotter till a situation called "meltdown" occurs, causing the theoretical "China Syndrome" where the core becomes so hot, it melts through the Earth and emerges in China – from a US perspective!

Of course there are other forces which we have used in the past, and are still using today – wind-power, water-power, muscle-power and other things like tension – like a clock spring or a rubber-band. Then we have used gravity too like a guillotine to chop someone's head off! But it is with the advent of steam that we really began to make some progress, and then electricity. Strangely enough, the gasoline engine has not been completely replaced and is still in use after more than a century, to drive our cars, as electricity has not completely taken over this role, due to the awkwardness of moving vehicles having to be linked to cables from the power source; batteries, though better than they were, are still not efficient enough. Then, chemical rockets to put our satellites into orbit still need to carry the weight of their fuel, which is a disadvantage.

We desperately need a new power source if we wish to make a further leap forward in progress. Our fossil fuels will run out someday, so the need to find another power source is very necessary. Nuclear fusion that happens in the sun is a possible answer, because unlike nuclear fission it doesn't give off so much dangerous radiation. It has been the dream of nuclear physicists to produce a fusion engine which they have not yet been able to do. But like all dreams, this may come true in the not too distant future.

But could there be other forces in nature that we simply know nothing about? For instance, present theories suggest that particles in giant clouds of nebulae clumped together to form larger and larger particles, eventually to become large enough to form stars and planets, under the influence of gravity. But consider would one speck of dust have enough gravity to attach itself to another speck of dust to form a larger speck of dust? So was it some other force that attracted the two particles to combine in the first place?

Performing experiments in 2002-2003 aboard the International Space Station, Mission Specialist, Donald Petit, conducted an experiment with various materials like coffee granules, and noticed that they tended to clump naturally together in the vacuum and zero gravity of outer space. He suspected the cause of the clumping could possibly be electrostatic. In other words the theory could be that the coffee granules first started clumping together under electrostatic influence, and then proposed that when the clump had achieved enough mass, gravity took over to continue the clumping. But could it be that there is some other mysterious force in space that repels matter causing it to coalesce? Gravity is the most pervasive force in nature, but that is with objects of enormous mass.

So what about the very tiny particles that have hardly any mass at all? It is difficult for scientists to perform such experiments under the influence of gravity on this planet. It is only in the zero-gravity of outer space, like the ISS that we could understand how particles behave.

[4]

Living Molecules

BEFORE LOOKING OVER OUR HEADS TO SEE WHAT MAKES THE UNIVERSE TICK, WE HAVE TO LOWER OUR HEADS AND TRY AND LOOK INTO OURSELVES TO SEE WHAT MAKES US TICK.

What we are made up of and how we developed from the very tiny to the very large. Basically we are a living colony of trillions and trillions of walking talking cells, the building blocks of all life on this planet, and not to be conceited, much more complicated than the simple hydrogen gas that makes up most of the Universe.

DNA

Molecules, made up of all sorts of different atoms, can become very, very complex indeed, particularly when the carbon atom, which is the basis of all life on this planet, links up with other atoms such as hydrogen, oxygen, nitrogen and phosphorus. Combinations of these atoms form nucleic acids which in turn arrange themselves to form the long DNA (deoxyribonucleic acid) molecule that you've heard so much about! All life has DNA.

DNA is in the nucleus of every cell of every plant, insect, fish, fowl or mammal. The nucleic acids which make up DNA are Adenine (A), Guanine (G), Thymine (T) and Cystosine (C), and these arrange themselves along the "double helix" DNA chain, which looks something like a twisted rope-ladder. The rungs on the ladder represent the chemical links (A, G, T, C) which hold the ladder together. When this ladder splits (imagine you have cut through all the rungs of the ladder, then pulled it apart like a zipper), free floating DNA units called nucleotides in the chemical soup of a cell, quickly latch on to replace these rungs and form two complete and separate ladders. This is how the DNA molecule makes a copy of itself. Once this has happened, the cell divides, taking one DNA ladder into itself.

The language of DNA is quite simple, really; just the four nucleic acids A, G, T and C. But it is the way that they are arranged that is so complex. It's something like the seven notes of music (and their respective sharps and flats) that, when put into a complex arrangement, become a Gershwin concerto!

Not only does DNA reproduce itself, it also carries the recipe for the manufacture of the all essential products called proteins. We generally think of proteins as muscle-building foods like meat and milk, but actually they are a versatile class of organic compounds made up of amino acids. Foods rich in protein, when broken down, give our cells the fuel for making their own proteins. There are countless varieties of proteins; hormones and enzymes are some prime examples.

PROTEIN PRODUCTION

How protein production happens is rather difficult to explain, but I'll give it a shot. What happens is that part of the DNA chain un-zips to expose the genes (long strings of groups of nucleic acids that are required to make a protein). Free floating molecules of RNA (ribonucleic acid, which is very much like DNA) in the chemical soup of the nucleus, quickly attach themselves to the exposed part of the DNA and make a copy of the gene in a "string" called messenger RNA in a three letter word language of A's, U's C's and G's. Notice the letter T in the DNA is replaced by the letter U in the RNA. That's a difference between DNA and RNA.

The string of RNA then detaches itself, while the DNA chain closes up again. This string of messenger RNA then leaves the cell's nucleus through a "porthole" in the nucleus' membrane, and enters the body of the cell to find a ribosome, which are "anvils" of protein production. The ribosome attaches itself to the messenger RNA string, and another type of RNA called transfer RNA is called up and reads the three letter "words" like ACU, CGG and CCU. These transfer RNA then bring the specified amino acid from the cell's sea of chemicals to the ribosome where the three letter words are matched up to the messenger RNA. In turn, the amino acid links on to a growing chain that will become the protein required.

All this sounds a bit complicated I know, but for a simplistic example it's a bit like going to a special kind of restaurant where meals are prepared according to your own recipe. You are the customer (DNA) and you want a special dish (the protein) just like "mother used to make."

The waiter (messenger RNA) comes up to you for your order. You tell him your mother's recipe (the gene), and he copies it down. He then goes out of the dining area (the cell's nucleus) and into the kitchen (the cell itself) and meets the chef (transfer RNA) standing near the stove (a ribosome). The waiter tells the chef what you want to eat and gives him the recipe he has written down. The chef then picks out all the ingredients (amino acids), required for the recipe and puts them together to prepare your specific meal (the protein). I hope all this makes sense.

CHROMOSOMES

DNA molecules come in tightly wound-up packages called chromosomes which lie in the heart of every cell. (Think of taking an elastic band and twisting, and twisting it between your fingers until you have a sort of "rubber band package"). Although these strands of DNA are very, very small (the entire DNA in all your cells would fit into a match-box), they are also very, very long.

If you unwound the entire DNA in your cells and strung them together in a single strand, it would reach the Moon and back over a thousand times!

GENES

Attached to the chromosomes on the long DNA string are, of course, the genes, and it is the way that these genes are arranged that spells out whether we develop into a human being, a cockroach or the geranium growing in your flower pot. It's a complex coded plan that not only decides what we're going to be, but also determines our hair, eye or skin colour, and our height, and our sex. A very simplistic way of looking at genes is likening them to a paragraph in a book. It is the arrangement of the words that makes sense and determines whether we wind up as "Lady Chatterley's Lover" or "Gone With the Wind"!

Each cell in our bodies carries our chromosomes in their nucleus; each cell that is, with the exception of the sex cells, the sperm and ovum. Normally a human cell has 46 chromosomes, but the sex cells only have half that number. When sperm and ovum meet, the normal complement of chromosomes is made up - half the amount of DNA from each parent. Hence, we inherit the characteristics of both our parents.

We are made up from billions and billions of cells. The various types make up our brain, heart, liver and lungs; our eyes, tongue, teeth and lips. We are just a walking, talking cell colony. Cells are tiny, and you could comfortably fit thousands into the period at the end of this sentence. The cells which make up our body are in a constant process of regeneration, which means they die and are quickly replaced, except perhaps our brain cells (neurons). As we get older, some cells die and are not replaced, and we can see evidence of this effect when our skin begins to wrinkle in old age. Strangely too, some cells continue living for a while, even after we eventually die, like our hair and fingernails, as any forensic pathologist knows.

[5]

Down from the Trees

SINGLE-CELLED ORGANISMS, LIKE BACTERIA, CONSIST OF ONE CELL ONLY, UNLIKE US WHO CONSIST OF ABOUT 100 TRILLION CELLS.

The earliest forms of life started out with single-celled organisms that gradually developed to multi-celled organisms, and eventually into us. A single-celled organism is surrounded by a membrane, contains DNA in its nucleus and has the necessary amino acids which the ribosomes use to produce the proteins necessary for its survival. We mustn't forget that at the point of conception, we were just a microscopic single cell ourselves, when the sperm and ovum combined, divided and began to grow. Cells are the building blocks of all life on this planet.

Unlike single-celled bacteria, viruses are not considered a life form by some, and are sometimes thought of as "organisms on the edge of life." They come in all sorts of weird shapes, and attack the cells of all living things – plants, animals and even bacteria.

They can only survive and reproduce within a living cell, sometimes mixing their DNA with that of the cell they inhabit. They cause all sorts of infections both in plants and animals, and the influenza and HIV virus are prime examples. There are many millions of types of virus, and they are so small (about one-hundredth the size of an average bacterium) and are so tiny they can't be seen under an optical microscope. But don't confuse these viruses with viruses that could infect your computer, as though they are both nasty, they are completely different things!

No one, not even the most brilliant scientist, can explain how exactly life began on Earth or where it came from. Some say it came from the soup of chemicals in the sea that developed into amino acids (the building blocks of life); some say it may have developed in volcanic vents under the sea; some say it may have come from comets or asteroids that seeded life on Earth (some evidence of amino acids on meteorites have been found). But no one yet really knows or can explain.

The history of our Earth goes back about 4½ billion years, when it began to form from all the debris lying around in the early Solar System. If we were to condense all that time into a single Earth year, then we could say from January to March things were very chaotic, and the Earth was a fiery molten ball with rocks crashing down on it from every direction. However, by spring, things began to settle down, and in about April, when the surface had cooled down sufficiently and oceans had formed, early life began in these oceans – first as single-celled creatures, then into more complicated aquatic forms like Trilobites by the month of May.

By about June, the continents were still dividing and by July the first plants began to form and grow. In August the seas were full of fish and by September the first insects appeared on the land. In October it was the reign of large reptilian creatures, the dinosaurs. But by the following month of November, they suddenly became extinct in some calamitous event. Most scientists agree that it was a comet or asteroid that crashed into Earth in the region of the Gulf of Mexico that wiped them out so suddenly, as all the fossils of dinosaurs appear in a narrow band of rock, which goes back 65 million years, called the K-T Band (between the Cretaceous and Tertiary period) – before this band no dinosaur fossils are found, and none afterwards.

This calamitous event caused the environment to change and wipe out most of the life on the planet, including plant life. So, most of the dinosaurs may have literally starved to death. However, some small furry shrew-like mammals apparently weathered the storm and were able to survive now that all the big guys had disappeared. They held sway and developed into larger mammals like sabre-toothed tigers and woolly mammoths. By about midday on the 31st of December, the last day of "our year" the first hominids (early ape-like man), began to stand up, and an hour before midnight Neanderthal man first appeared. By quarter-to-midnight, homo-sapiens (us) lived in caves and began to make weapons.

The history of modern man goes back to only 5 thousand years which is within the last few seconds of "our year." Before that we can't trace our history as we were not civilised enough to or develop any sort of written language.

As Darwin concluded in his Origin of Species, all life evolved from species to species, improving on itself in order to best survive. Eventually man evolved to the top of the tree of life from early primates (a bit ironical that, knowing primates can climb trees better than we can!) But there is simply a huge difference between primates, like chimps, and us. There is nothing in-between man and chimp – the so-called "missing link". Loveable as chimps are, there is no match between their intelligence and ours.

For example a child plucked from even the most primitive human tribe on this Earth has the intelligence to match that of a Cambridge graduate, given the right environment and education. In other words, he has the capacity to learn and achieve the highest degree of intelligence. A chimp hasn't; once a chimp, always a chimp. Why is this great gap between us and our nearest biological relative? There are even wild theories that early alien visitors to our planet, mated with early man (or should I say woman), and mixed their DNA with ours, creating an intelligent species. Dismiss this if you must, but we are still searching for an answer to this perplexing question of why we are so different from the animals around us.

Size-wise we are not doing too badly either, with only a handful of land animals larger than us. Life comes in all forms, shapes and sizes – from microscopic bacteria to giant sequoia trees. It is essential that we are of a reasonable size to enable us to support and carry our large brains (we have the largest) whereas large animals need only be big enough to support their large stomachs!

Also, we walk upright, unlike other animals, which is a great advantage, as we first started to show signs of intelligence when we began to stand up. The dextrous human hand too has helped us develop into what we are, as it is so much more efficient than say, a cat's paw. So, you could say that hand and brain have made us the intelligent and advanced species that we are.

How long man will continue to dominate this planet is up to us. For the first time, we have the ability to destroy our world in a nuclear holocaust, or destroy our environment with pollution. Outside events like ice ages (we have had more than one in our earlier history) or the cataclysmic event of a large comet or asteroid crashing into us could wipe us out too or send us back to the Stone Age! However, our little planet should continue to revolve around the Sun for billions of years, till the Sun eventually swallows it up. If we were to suddenly disappear from the face of the Earth, other creatures may evolve to dominate the planet, and within just a few thousands of years, all evidence that we humans existed at all will be eradicated – our great cities would crumble and decay, to finally disappear altogether.

However, to be more optimistic, we may go on for several thousand years, and with the advancement in technology, may finally reach for the stars! In the meanwhile, in time to come, we Earthlings must colonise at least Mars and maybe some of the more friendly moons of Jupiter and Saturn in order to ensure our survival.

I hope in the foregoing I have been able to give the reader a brief story of life – how it has progressed from atom to molecule, from molecule to cell, from cell to us. It is the recipe for all life on this planet, and probably the most complex in the entire Universe.

[6]

Over Our Heads

TO OUR NORMAL SENSES THE UNIVERSE IS A VERY REMOTE PLACE INDEED. EXCEPT FOR THE SUN WHOSE HEAT WE FEEL ON OUR SKIN, AND THE MOON, THAT MIGHT AFFECT OUR MOODS, IT IS ONLY FROM SIGHT THAT WE CAN SENSE THE REST OF THE UNIVERSE.

Our other senses of hearing, smell, touch and taste don't play any part in our awareness of this great canopy over our heads. Of course, astrologers always believed that it was the position of heavenly bodies, the Sun, Moon and planets and where they were placed among the constellations of the background stars that influenced our lives and fate. This was very significant to anyone who believed in their horoscope, and is even true to this very day. But seeing the Universe is not quite enough. One has to "think" the Universe to become aware of its wonders, and it is only with our brain that we, a colony of cells and DNA, can try and comprehend something so big and wonderful.

Looking up at the stars, the night sky is one of nature's most beautiful sights, and one of the most mysterious if you realise that you are looking back through time in hundreds, thousands and even millions of years. For instance, a fuzzy patch of light is visible to the naked eye in the constellation of Andromeda. This is our neighbouring galaxy and is so far away, that you are looking at it not as it exists today, but as it existed over 2 million years ago, when early ape-like men roamed the prairies of Africa! If you have keen eyesight and the night is clear, you will probably be able to see about 2,000 stars - some will be bright, some dim, and some smudgy. You may also notice that some are dazzling white, while others have a bluish or reddish tinge.

It is worth observing the night sky on a clear night, and you don't even need any astronomical paraphernalia to do this. If possible, borrow a book of night sky maps from your local library, and try to get your bearings. Some groups of stars seem to make patterns in the sky, and in ancient times, observers gave these the names of their gods and mythical beasts. Although it is hard to imagine these shapes, you will at least be able to pick out the constellation of Ursa Major (the Great Bear). In the United States it is more popularly known as the Big Dipper because it does resemble that shape.

Cassiopeia is also quite easy to see, just opposite Ursa Major. It is like a big "W" in the sky. Soon you will become more familiar with the other constellations, like Cygnus the Swan and Orion the Hunter, and you will soon notice that you can only see some constellations at particular times of the year; others you can see all year round.

Three factors determine how the sky looks at night; the time of night, the time of year and your latitude (how far north or south of the equator you are) on Earth. The last is the most confusing and difficult to explain. From our viewpoint in the British Isles, you will notice that all the stars gradually revolve around a point in the sky where the Pole Star (Polaris) is located. To gain the same sort of effect, imagine you are pointing an open umbrella in the direction of the Pole Star. Imagine the umbrella has some of the northern star constellations painted on the inside. Now if you gently twirl the umbrella in a counter-clockwise direction, the constellations would imitate the way in which the night sky revolves around Polaris. Depending where you live on Earth, the Pole Star would vary from directly over your head (as it would be seen from the North Pole), to just above the northern horizon (near the Equator) if you are living in that region; so the angle that you point your umbrella would have to vary according to your latitude on Earth. In the southern Hemisphere, you would not see the Pole Star at all!

Sometimes the night sky puts on a great display when the Earth passes through the remnants of gas and dust trails left by past comets, when a shower of meteors, or shooting stars, encounter the Earth's atmosphere, and burns up. For example the Perseids are an abundant shower of bright meteors that emanate from the region of Perseus every August. If you are lucky enough to witness such an event in the clear night sky, you will notice that these showers seem to radiate from a small point of the sky, so for example if they emanate from the region of the constellation Lyre, they are called Lyrids, and if from the region of Gemini, the Geminids and so on.

If you are fortunate enough to own a pair of binoculars or a small telescope, the wonders of the night sky will open up to you even more, and you will be able to see further into the Universe and observe its wonders. Stars, themselves only display a spangle of light (you can never see the disk of the star itself, even in the largest telescope). Nebulae is far more interesting and it's great fun to focus on what you think is a fuzzy star, only to discover that it is really made up of several stars lighting up a patch of nebula, like the great Orion nebula.

It is also possible to split what appears to be a single star to the naked eye, and find it is really two stars – a binary system, For example the double star, Albireo, (in the constellation of Cygnus the Swan) is one of the sky's showpiece doubles, consisting of a yellow supergiant star and a blue-green companion. The splitting of a double, like Mizar and Alcor in the panhandle of the Big Dipper (the Great Bear or Ursa Major), can sometimes be done without the aid of a telescope, if you have very keen eyesight.

If you are beginning in astronomy, I would suggest a good pair of binoculars with a wide field. You will be amazed at how the Universe opens up to you with the help of binoculars, and you will see many, many more stars than you did with your naked eyes. They are reasonably cheap and easy to handle, and don't require any setting up as some telescopes do.

Telescopes are more expensive, and to invest in one requires careful consideration. There are two basic types, refractors which is a tube with a lens at one end (the objective) and an eyepiece at the other.

It is the sort of telescope that Galileo used. The other type is a reflector which uses a combination of mirror (to collect the light), lens (the eyepiece) and prism. Sometimes these are known as Newtonians, as they were invented by none other than the great Sir Isaac Newton himself! The objective lens in a refractor and mirror in a reflector, determine how powerful the instrument is going to be. In other words, the larger the lens or mirror, the more light it can gather. Hence, refractors are somewhat limited by how large a lens can be accommodated on the equipment.

Mirrors on the other hand, can go to enormous sizes – from 4 inches across, for a very small mirror, to monsters of over 300 inches for a really large one. The Hubble Space Telescope uses a mirror of over 94 inches. Now segmented mirrors are being used, like in the Keck telescope, to go beyond the limit of a single 300 inch mirror, which is technically difficult to build and very, very expensive. This array of smaller mirrors act together as a single big mirror, and can go up to almost 400 inches. For a serious aspiring astronomer, a 6 or 8 inch mirror will reveal most of the objects in the Universe. More compact but expensive varieties of telescopes are the Schmidt-Cassegrain series.

All telescopes require a sturdy mount, and the most popular are the altazimuth and Dobsonian mounts, but the ideal mount for tracking a star (you'd be surprised how quickly they move in the telescope's field) is an equatorial mount. This has to be set up, but once done it tracks the object in the sky keeping it in view as it moves. Modern telescopes now have built in computer controls that not only find a particular object in the night sky but track it as well. The sky's the limit as far as modern telescopes are concerned, and one can only imagine what problems early astronomers like Herschel encountered in their efforts at star-gazing.

Besides your binoculars or telescopes, you need a really good book of star maps or at least a chart like a "Planisphere" which is on a moveable disk, that shows the night sky and how it looks at a particular hour, on a particular date of a particular month. You may notice on your star charts that many of the stars are labelled with a letter of the Greek alphabet. The brightest star in a constellation is usually (but not always) assigned the Greek letter α (alpha), moving down in order of brightness to β (beta), γ (gamma), δ (delta), ε (epsilon), η (eta), and so on.

Your book of star maps will probably give you the entire Greek alphabet. You will also notice that some objects are given M numbers, like M1 (Crab nebula), M8 (Lagoon nebula), M16 (Eagle nebula) which were originally given by the French astronomer, Messier in an effort to distinguish these fuzzy objects from comets. They are still in use today, but have now been given their New General Catalogue of Deep Sky Objects - numbers like NGC 1952 (the Crab), NGC 2563 (the Lagoon) and NGC 6611 (the Eagle).

Your star maps will also indicate the apparent brightness (how bright it appears to us) of any particular star. The brightest star in the sky is Sirius at magnitude -1.4 and the dimmest, or faintest stars that can be seen with the naked eye, at magnitude 6 (the higher the number, the fainter the star).

Stars with a magnitude of over 6 can only be viewed by binoculars or a telescope.

I hope I have armed you with sufficient ammunition to whet your appetite for braving the weather, and on the next starry night venturing out in to your back garden to gaze up at the starry night sky over your head. Happy star gazing!

[7]

Star Worshippers

EVER SINCE HE FIRST STEPPED OUT OF HIS CAVE, MAN HAS LOOKED UP AT THE STAR-SPANGLED NIGHT SKY AND WONDERED ABOUT THE MYRIADS OF LIGHTS TWINKLING AND SPARKLING THERE LIKE JEWELS.

Today, aeons later, astronomers who probe the Universe with powerful telescopes and other apparatus, are still mystified and intrigued by some of the objects revealed at the very fringes of space and time. It would seem that the Universe is still jealously hanging on to some of its secrets!

ANCIENT CIVILISATIONS

Like the ancients who thought we were at the centre of the Universe, we'll start at the very beginning with planet Earth itself. To them, the motions of the Sun and Moon "around the Earth" were of the utmost importance (and they still are!) The Sun progressed across the sky from east to west, describing an arc. Initially, the ancients thought the world was flat and was enclosed by a sort of "celestial dome" where the Sun predominated by day and the Moon by night. The dark night sky was also studded with pinpoints of light that peeked through. This was thought to be the "heavenly fires" that blazed behind the celestial dome of the sky. The most important events in the day was the rising and setting of the Sun. The ancients were able to measure time by dividing this cycle into two parts - night and day. But to confuse them, not only did the Sun describe a gentle arc in the heavens (the ecliptic), they also noticed this arc throughout the year, climbed higher in the sky each day until it got to a certain point, then it started descending to the lowest point, before ascending again.

They associated the high peak with "summer" and the low with "winter." It told them the best time to plant seeds and when to gather their crops. In some places like ancient Britain as far back as 3,000 years before Christ, they even went to the extreme of hauling huge stones half-way across the island to create megaliths like Stonehenge. This prehistoric "astronomical observatory" was oriented towards the position where the Sun rose to the highest point of its arc - the summer solstice. Another 500 years later, the Ancient Britons, were probably able to predict eclipses of the Moon!

THE FIRST OBSERVATORIES

The ancient Egyptians too must have been very aware of celestial motions, but they were more concerned with how the flooding of the Nile was associated with the first visible rising of the star Sirius. So the early Egyptians had rather a limited interest in astronomy. They even gave up attempts at working out a calendar because they found it all too complicated. However, according to recent theories, they did attempt to "paint" a map of the constellation of Orion on the desert, building the three Pyramids of Giza to represent the three stars in the belt of Orion, with that "river of stars," the Milky Way, representing the Nile. The Babylonians, on the other hand, were much more interested in matters astronomical, and by about 1800 to 400 BC developed a calendar and worked out the motions of the Sun and phases of the Moon. They were good at their maths, and were able to predict quite accurately when the next new Moon would appear.

SPHERES OF INFLUENCE

The Greeks came next, though they used a geometrical rather than a numerical approach to gain an understanding of the motions of celestial bodies. Although by the time, they began to abandon the theory that the Earth was flat, they still held on to the idea of an Earth sitting in the centre of the Universe with the Sun, Moon, planets and all the stars whirling around it, fixed in "concentric crystalline spheres". One can scarcely blame them for this misconception. After all, "seeing is believing" Or is it? How often have you been fooled into thinking your train is moving when it is actually the one standing alongside which has started off? Seeing can be deceptive, if you rely on your eyes alone.

However, unlike most of the stars, which were firmly fixed in their places in some sort of "celestial backdrop," a few of them would not conform to their neatly ordered Universe. Firstly these followed the path of the Sun, which the other stars didn't do, and secondly they had the annoying habit of appearing to advance against the background of stars for a while, then erratically turn back on themselves in a kind of loop, before advancing again. The early Greeks were mystified and called these odd objects "wanderers" or "planets" as we still refer to them to this day.

Another irritating feature of these planets, which was difficult to explain, is why they got smaller and dimmer at different times of the year. To fit in with these strange movements, the Greeks had to modify their model of their concentric ring universe, and include little circles (epicycles) - for each planet - that moved round the larger circle (the deferent). They also had to place the Earth slightly off-centre in the Universe to get their model to agree with actual observations of the celestial bodies. It was the case of back to the drawing board each time some discrepancy occurred, and it was finally Claudius Ptolemy who tidied up all the irregularities and produced his Almagest, around about 127 AD. This huge compilation was the "Bible" of astronomical knowledge up to that time, and besides Ptolemy's own ideas, it included the thoughts of other famous Greek philosophers like Aristotle and Hipparchus. This great work eventually passed on to the Islamic and the European civilisations.

So, now happy with their explanations, the Greeks rested their case. And it rested for a millennia-and-a-half, till Copernicus came on the scene!

To the ancient Greeks, the heavens were the preserve of the gods and mythological heroes, not to be interfered with by mere mortals. They connected the stars with imaginary lines to create constellations in the night sky - shapes like Pegasus the Winged Horse, Orion the Hunter, Scorpio the Scorpion and Cygnus the Swan. For example Orion is depicted as the figure of a man holding a shield in one hand towards the snorting bull,

Taurus, and brandishing a club with his other. Following at his heels are his two dogs, Canis Major and Canis Minor. Also, according to legend, the boastful Orion was supposed to have been stung by a scorpion at one time or other so, prudently, always kept well away from this stinger, setting in the sky before Scorpio rose. Great epic tales were woven about these legendary heroes and all their exploits.

Another important purpose the stars and planets fulfilled was the prediction of earthly and human events. The arc of the ecliptic was divided into twelve zones called The Zodiac, and depending on the position of the Sun, Moon and planets within these zones, horoscopes could be written. The appearance of a comet was considered a prediction of terrible doom, and an eclipse was another fateful omen. It was to work out these events and the position of the heavenly bodies with advanced mathematics that led to the preparation of early calendars.

The Greeks obvious interest in astronomy is borne out by the discovery, in the early 1900s, of an astronomical calculator made of bronze and using gears which was an amazingly complex instrument for an ancient culture to produce. It was recovered from the wreck of an ancient Greek ship, and is called the Antikythera Mechanism. The device is extremely old and a very rare find. It is thought to have been made in the 1st Century BC. Even then, it seems to have been way ahead of its time and nothing like it has been found till early astronomical clocks of the 14th century. When it was first discovered it was not clearly understood what it was, and it is only now, over a hundred years later, that we can understand the complexity of its construction and for what purpose it was designed.

So, although our early ancestors may not have understood the Universe, they certainly took the time to look over their heads and wonder about it. Under clear Grecian skies the night sky must have looked spectacular, with stars glittering like gems scattered on a background of black velvet.

We, poor unfortunates, living near cities with all the light pollution, are lucky even to be able to see a few of the brighter stars, let alone the splendour of the Milky Way.

[8]

Star Gazers

FOR AEONS OUR EARLY ANCESTORS HAVE GAZED INTO THE HEAVENS AND OBSERVED THE STARRY NIGHT SKY AND THEY ALWAYS THOUGHT THAT IT WAS THE DWELLING PLACE OF GODS.

As a result, they never fully comprehended what they were really seeing. It was only when we began to think about the Universe that we began to wonder what it was all about, and how it worked. To think philosophically was always considered heresy and conflicted with religious thought. The Earth was always considered central to the Universe, and in Biblical terms was made before the stars in the firmament (the heavens).

It was only with the advent of the early European astronomers in the 16th Century, that these religious beliefs were challenged, and astronomical studies really began to take off. The most important of these brave pioneers of scientific thought were:-

COPERNICUS

In 1514 Nicolaus Copernicus entered the tidy and stable model of the "Ptolemaic Universe." While carrying out his ecclesiastical administrations in the Roman Catholic Church at Frauenburgh, Poland, he had been quietly flirting with his first love, astronomy, studying the movements of the stars and planets. He wrote down all his thoughts in a massive thesis saying that it wasn't heavenly bodies like the Sun and planets that rotated around the Earth, but just the opposite; Earth and all those wandering planets went around the Sun instead! However, the concept of solid celestial spheres and the perfect circular motion of heavenly bodies were incorporated into his model, except that the Sun now took the place of the Earth at the centre of the Universe.

This nearly solved all the complicated movement of celestial bodies that Ptolemy and his predecessors had sweated over. But you can just imagine what consternation this revolutionary theory might have caused, especially to the Church where Copernicus served. It moved the centre of the Universe away from the all-important Earth, which, since Biblical times held pride of place in the middle of the Universe. To think otherwise was heresy!

You can understand why they (the Church) would have been upset. It was the wrong time to come up with any new ideas. The Church dominated just about everything and everyone. Although his radical ideas leaked out to most of Europe, Copernicus could not very well broadcast this to the world at large and "boldly go where no man had dared to go before." So in the end he patiently played the waiting game and delayed the publication of his theories. Meanwhile, probably to appease the Church, he went along with the Ptolemaic theory. It is generally believed that he was eventually able to see a printed copy of his treatise, albeit on his death-bed in 1543.

TYCHO BRAHE AND KEPLER

Following in the steps of Copernicus came Tycho Brahe and Kepler. Tycho worked as Imperial Mathematician to the Holy Roman Emperor, Rudolph II. Kepler came to work as Tycho's assistant, and when Tycho died in 1601, all his observations and calculations passed to Kepler. He used these to good effect, but still could not make his calculations work out in relation to his actual observations regarding the movements of the planet Mars, using the Copernican model of perfectly circular planetary orbits. At last he came to the conclusion that the orbit of Mars was not circular but elliptical. Using this principle, he was able to explain the movements not only of Mars, but also the orbits of the other planets, including Earth. He published his "Laws of Planetary Motion" between 1609 - 1619. Copernicus' model of the Universe and Kepler's ideas of elliptical orbits for the Earth and other planets, helped greatly in calculating much more accurately planetary distances in relation to the Sun, thus giving astronomers, at last, a clearer picture of the Universe as far as the movements of the planets around the Sun were concerned.

GALILEO

While Kepler was publishing his first two Laws of Planetary Motion, Galileo Galilei, an Italian mathematician, was gazing up at the night sky through a new-fangled instrument he'd designed (but not invented) called a telescope.

Although it was primitive by today's standards, it was quite fantastic to Galileo. He made many observations of the planets, and published papers on his findings. Of particular interest were the four moons of Jupiter, which he could see through his telescope as tiny pinpoints of light. They danced around the planet behaving like a miniature Solar System, and he used this argument to prove that at least these moons, did not pay homage to the Earth by circling around it, but rather to Jupiter! He also studied the planet Venus and noticed it had similar phases to the Moon. But he accepted that the Moon was the only heavenly body that did revolve around the Earth!

Like Copernicus, Galileo reckoned from his eight months of observations that the old picture of the Universe (the theory that the Earth was at the centre) was incorrect; Aristotle and Ptolemy were mistaken, and even the book of Genesis was wrong! Galileo was like a disruptive comet suddenly appearing in orderly skies! The Church authorities became extremely upset by his revelations. He tried hard to persuade them - probably even allowing them a peek or two through his telescope - they were not convinced. They stubbornly maintained that he must forget his belief in the Copernican system and stick to what the established Church believed in - that Earth was central to the Universe - and not stir things up. Galileo, however, was rather an abrasive sort of chap, and he really got into the Church's bad books. So he became an excellent candidate for the dreaded Inquisition. To avoid rotting in prison and enduring the terrible tortures that could be inflicted on those whom the Church considered heretics, Galileo eventually threw in the towel and was forced to admit to the "error of his ways."

Who can blame him? After all, it was hardly worth being subjected to the pain of the rack, thumbscrew or worse, simply in the name of Science! Because Galileo was willing to capitulate, his inquisitors relented and his prison sentence was swiftly commuted to house-arrest. So at least he was able to pursue other scientific research in comparative peace, until his death in 1642. (Dogmatically, the Church still held on to its pre-Galilean concept of the Universe, and only as recently as 1993 did the Vatican officially admit that Galileo had been right all along!)

NEWTON

Next on the scene came one of the giants of scientific thinking, Isaac Newton. As the bubonic plague ravaged England in 1665, Newton, who was 22 at the time and studying in Cambridge, went home to carry on his studies. Tradition has it that while sitting in his garden underneath a fruit tree, he watched an apple falling to the ground. It's generally known that this is just a myth, but it's a romantic myth, so let's leave it in.

He decided the Earth had exerted some sort of force on the apple, causing it to fall downwards towards it. He called this force "gravity." Going beyond the apple, he reckoned that even the celestial bodies attracted each other with this gravitational force. He published his findings on the laws of motion, which contained the familiar phrase: "To every action there is always an equal and opposite, or contrary, reaction." His theories explained the Moon's gravitational effect on tides and, going one step further than Galileo, he made the first reflecting telescope – one that uses a combination of lens and mirror.

It is still the basic design for some of the largest telescopes in operation today. In 1705 Newton was knighted by Queen Anne. He died in 1727 at the ripe old age of 84, and was buried in a monument in Westminster Abbey.

HERSCHEL

Following on the heels of Newton, came the German-born musician William Herschel. He escaped from Germany in 1757 at the age of 19 because Hanover at that time was under occupation by the French. He made a living for himself in England for the next 25 years or so as a professional musician, but his first real love was astronomy. He became interested not only in astronomy, but also in the making of telescopes, and constructed the largest one ever built up to that time. It was forty feet in length, of poor optical quality and rather cumbersome and unwieldy, which meant it could be manipulated only with the greatest of difficulty. Herschel's most important contribution to astronomy was the discovery of the sixth planet, Uranus, which was not visible to the naked eye.

Herschel also detected double-stars with interlocking orbits, thus proving that Newton's laws on gravity extended beyond the Solar System. Not only that, he also discovered the general shape of our Galaxy, the Milky Way, in which our Sun lives, and the course it follows through space, carrying with it the Earth and other planets. Following family tradition, his sister Caroline discovered eight comets, and his son John also became a great astronomer in his own right.

MESSIER

Another famous astronomer at around the same time was Charles Messier, a Frenchman, who published a catalogue of stars in 1774. He was mainly interested in the discovery of new comets and, in order not to confuse them with other fuzzy objects in the night sky, he gave these other nebulous objects "M" numbers, which are still in use today to identify such beautiful things as the Crab Nebula (M1), the great Orion Nebula (M42) and the Andromeda galaxy (M31).

Besides producing this famous and still-used list of celestial objects, Messier was an astute observer, studying eclipses, occulations and sunspots, and was the first astronomer in France to observe Halley's Comet on its return in 1759.

[9]

Star Chasers

THE ANCIENT SCIENCE OF ASTRONOMY WAS MORE CONCERNED WITH THE DIRECT OBSERVATION AND MOVEMENT OF CELESTIAL OBJECTS.

By the 19th and 20th centuries, this science blossomed out into the realms of astrophysics which used a more scientific base to explain observed celestial phenomena and physical processes.

With the dawning of the 20th century, this branch of science really took off in leaps and bounds, and our understanding of the Universe today is largely due to the work of some amazing scientific geniuses.

Undoubtedly, the geniuses of the 21st century are already learning the basics based on these predecessors so they in turn will expand our understanding of what's over our heads.

EINSTEIN

Albert Einstein was born in Ulm, Germany in 1879, of German-Jewish parents. They moved to Munich when Albert was an infant, and set up a small electrical and engineering business there. But when this failed, the family decided to leave Germany and live permanently in Italy when Albert was about 15 years old. After failing an exam that would have led to him becoming an electrical engineer, he spent the next few years in Zurich studying to become a teacher in mathematics and physics. However he wound up, a couple of years later, in a Swiss patent office and while employed there, wrote an astonishing range of publications in theoretical physics, largely in his spare time.

He obtained his Ph.D. at the University of Zurich in 1905, and after that received a regular appointment as associate professor of physics in the same university. By the age of 30 he was recognised throughout German-speaking Europe as a leading scientific thinker, and in 1914 was director of the Kaiser Wilhelm Physical Institute in Berlin. He remained on the staff until 1933, when, with the rise of fascism, he moved to the United States and became an American citizen.

Einstein's General and Special Theories on Relativity were revolutionary to the scientific world. He held that motion, time and distance are not "absolute" but "relative" to moving frames of reference, and he explained the Universe in terms of curved space and time, incorporating acceleration and gravitation into his concepts. His theory that gravity bends light was tested and verified during an eclipse of the Sun in 1919, and from that day on, Relativity became an everyday term.

His equation E=mc2 (energy equals mass times the speed of light squared) opened to mankind the awesome power of the atom. Though a pacifist and an outspoken supporter of world peace, Einstein suspected that Germany was conducting research into nuclear physics with the intention of producing some sort of nuclear device, and was persuaded to write to President Roosevelt warning him of that possibility. This helped influence the setting up of the Manhattan Project (research work led by Robert Oppenheimer) and the creation of the atom bomb.

Einstein, together with Galileo and Newton, stands out as one of the great conceptual revisers of our understanding of the Universe, and the way in which we look at space and time. By the time of his death in 1955, this gentle physicist with the familiar tousled mane had won the world's reverence for transforming man's understanding of the Universe.

HUBBLE

Our very first concept of a Universe, encompassing all celestial bodies in the starlit sky, was thought to be part and parcel of our own Galaxy, the Milky Way. Edwin Hubble changed all that. Born in Missouri in 1889, he graduated at the University of Chicago where he studied physics and astronomy. He also obtained a law degree at Oxford and, amongst other things, was also an excellent heavyweight boxer! After the end of World War 2, Hubble joined the Mount Wilson Observatory.

It was here that he had the chance of observing the Universe through the new 100 inch telescope sited there. He argued that a mysterious spiral nebula in the constellation of

Andromeda was not, as was hitherto thought, a cloud of gas and dust in our Galaxy, but was in fact another galaxy just like, and outside, our own.

This concept of "island universes" pushed back the boundaries of the Universe till it was much, much larger than anyone had ever conceived. Since then, hundreds upon thousands upon millions of other galaxies have been observed and studied, and our whole concept of the Universe changed forever. Hubble also observed that the whole Universe is expanding, establishing a ratio between the galaxies' speed of movement and their distance, which became known as "The Hubble constant."

Edwin Hubble died in 1953 and, as a fitting tribute to his memory and his contribution towards our understanding the Universe, the Hubble Space Telescope has been named after him.

CARL SAGAN

Those of you who watched TV scientific documentaries in the 1990s would remember Carl Sagan, the eminent American astronomer, astrophysicist and cosmologist with his somewhat Kermit-the-frog-like-voice, who featured in so many of these programs.

Most of his life was spent as a professor of astronomy in the Cornell University where he taught. He was a great supporter of the SETI (Search for Extra Terrestrial Intelligence) project, and among other books and papers, wrote the excellent book "Contact."

Some of you may have seen the film made in 1997, which sadly he didn't live to see, as he died a year earlier aged only 62.

STEPHEN HAWKING

If ever there were a more tragic case of a brilliant mind being trapped in a physically crippled body it is true of Stephen Hawking, the world-famous English physicist, cosmologist and scientific thinker. Among his many books, his "Brief History of Time" stayed on The Sunday Times as a best-seller for a record breaking 237 weeks. He has attempted to equate the theories of general relativity with quantum mechanics, which even Einstein baulked at, and he is a firm supporter of "The Many Worlds Interpretation" of quantum mechanics, which I have touched on briefly in this book.

His speech-generating device will be all familiar to those who have watched a scientific documentary in which he has featured, as he is almost completely paralysed by motor neuron disease which he has suffered from, and which has progressed through the years. His theories and predictions that black holes emit radiation is sometimes referred to as Hawking radiation. Stephen Hawking lives in Cambridge, England, where he has lived, studied, and worked for most of his 70 year life.

[10]

The Solar System

TO UNDERSTAND OUR PLACE IN THE COSMOS, THE BEST PLACE TO START IS IN OUR COSMIC BACKYARD. HERE'S WHERE WE CAN BEGIN TO GET A SENSE OF THE SCALE OF EVERYTHING.

The Solar System comprises the Sun, with its family of nine planets (now officially eight since poor little Pluto has been taken out of the equation and de-classified as a planet) all circling around it in their respective orbits. Some of the planets in turn have moons that orbit around their mother planets, and except for asteroids - big rocks really - and the occasional visiting comet, that's it - the Solar System.

With the exception of the planets, the odd satellite or perhaps a passing plane, all those other shining points of brilliant, sparkling lights you see in the night sky with your naked eyes, are all stars, and don't belong to the Solar System. Another important point to bear in mind is that the Sun and the stars are the only objects that shine with their own light. The Moon and all the planets only shine with the reflected light of the Sun.

Planets are sometimes hard to distinguish from the myriads of stars, and there are only five of them that you can see with the naked eye. If you familiarise yourself with the night sky, the easiest way of finding a planet is to look around the ecliptic (the apparent path the Sun follows in its journey through the sky) and determine whether there is a bright object that disturbs the usual pattern of star constellations. Astrologers are very aware of the movement of the planets through the zodiacal signs, as they consider the appearance of certain planets in the associated constellations have an effect on your life and fortunes!

THE SUN

The most obvious object in the Solar System, is of course, the Sun. To us it is the most important thing in the Universe. Without it there would be no life on Earth. On the face of it perhaps we could do without the Moon, if need be; perhaps we could do without the planets. The stars may be the easiest to dispense with. They are so far away that it may not matter if there were no stars in the Universe. From a purely selfish point of view, we only need the Sun and the Earth, and if these were the only two bodies in the Universe, we could still live, eat and breathe.

The Sun is just an enormous ball of glowing, fiery and boiling gas, caused by nuclear reactions deep in its core. It is made up mostly of hydrogen (just under three-quarters) and helium (just over a quarter), the two lightest elements. The nuclear process which causes the Sun to burn, is called nuclear fusion, which converts the hydrogen to helium, giving off a huge amount of energy in the process, and it's the same reaction that is produced by a hydrogen bomb, but in a controlled way. The Sun is so large that you would have to line up over one hundred Earths to cross its disc; its surface temperature (5,800 degrees C) is hot enough to boil metal and its internal temperature, at its core, is an incredible 15 million degrees Centigrade; it is so far away (about 93 million miles) that it would take you about 20 years if it were possible to get there by cruising in a jet airliner, whereas at the speed of light, it is just about 8 minutes away.

FORMATION OF THE SOLAR SYSTEM

Today's scientists reckon that the Sun and all planetary bodies, including the moons, formed out of an immense primeval nebula (or cloud) of interstellar gas, dust, and other matter that coalesced into lumps, the largest of which was the beginnings of the Sun. These lumps of matter, gradually condensed under the influence of gravity, and in the case of the Sun, the pressures and heat built up to such an extent, that nuclear reactions began at the core. This caused it to "ignite." providing heat, light and energy. In the case of the planets, there was not enough matter to cause this reaction, (though scientists say that Jupiter came close to it and it too might have "ignited" and become another Sun).

Previously, the Solar System was thought be quite unique (or at least rare) in having planets orbiting its central sun. But recently extra-solar planets have been detected using various methods. We can now detect whether the star wobbles slightly, like a hammer-thrower wobbling as he swings. It is impossible to see the planet directly, as its feeble light would be lost in the glare of the star – like a firefly circling round a searchlight. The planets that have been discovered this way are strange, very large Jupiter-like objects, circling close to their mother star. Another detection method is to measure the dimming of light from the star. This method can even be used to find smaller, rocky inner planets.

Note that most other star systems consist of two or more suns, orbiting around a common point, and it is unlikely that planets could form in this unstable environment.

SCALING DOWN THE SOLAR SYSTEM

To get an idea of the comparative size of the Sun and planets, and also the size of the Solar System, we need to scale it down millions and millions of times to a measurable size. Distances in miles are so great and the figures so large, that they hardly mean anything to anyone in everyday terms. Astronomers sometimes use a measurement called the "Astronomical Unit" when talking about the Solar System. An Astronomical Unit, or AU, is the distance of the Earth from the Sun (93 million miles).

If we were to scale the Solar System down so that one mile would represent an Astronomical Unit, and place it over the map of a well-known city like London, we can get an idea of the relative size of the Solar System. Suppose we were to use the London Eye as the centre, and place the planets in familiar parts of the City. Imagine our Sun as a huge hot-air balloon about 36 feet across, hovering in the sky above the London Eye. On this scale, the planet Mercury would be about the size of a plum, and lie in the vicinity of Trafalgar Square. Venus, the size of a grapefruit would be in the region of Aldwych, and we ourselves, again like a grapefruit, would be in Piccadilly Circus.

Further out, and about the size of a mandarin orange, would be Mars hovering around the dome of St. Paul's. Jupiter, a huge beach ball standing 5 foot tall, would be in Shepherd's Bush, and Saturn a large golden balloon measuring over 4 feet across, would be somewhere in Kingston-on-Thames. In the region of Leatherhead, Uranus would float, a large balloon of over 1½ feet across, and further away in Farnborough would be Neptune, almost the same size as Uranus. In the vicinity of Reading would be remote Pluto, the size of a pea on the very fringe of the Solar System.

Hopefully the foregoing will give you an idea of just how large the Solar System really is. There is a lot of empty space in it too, just to ease your mind that a comet or asteroid might suddenly come crashing into us! It's no wonder that a comet impact with Earth would be a rare (though catastrophic) event if it happened, or ever were to happen. But one can take comfort at the thought, after viewing a doom and gloom film or documentary of such an event, that such an occurrence is extremely rare. There have only been a handful of these visitors that came close enough to the Sun to be observed during the last 76 years, and after all, the comet or asteroid could just as easily choose Mars or Venus as its target instead of us!

I've tried to draw this model of the Solar System roughly to scale, because it's difficult for most illustrations in books to give you an appreciation of its sheer size, and also to relate it to a well-known city.

You could have fun using this 'scaled-down formula' for your particular town or city by choosing a central point of reference, (where you want to locate your sun), and placing the planets around it at roughly the following distances. This is so easy to do now with facilities like Google Earth maps.

MERCURY – UNDER HALF-A-MILE
VENUS – ALMOST THREE-QUARTERS OF A MILE
OUR EARTH – A MILE
MARS – JUST OVER A MILE-AND-A-HALF
JUPITER – ABOUT 5 MILES
SATURN – ABOUT NINE-AND-A-HALF MILES
URANUS – ABOUT 19 MILES
NEPTUNE – ABOUT 30 MILES
PLUTO – ABOUT 39 MILES (ON AVERAGE)

These 'cosmic miles' by the way are more commonly referred to as Astronomical Units (AU) so the Earth is 1 AU from the Sun.

[11]

Planetary Formation

EXACTLY HOW PLANETS FORMED ROUND THE SOLAR SYSTEM IS SOMETHING THAT SCIENTISTS ARE STILL TRYING TO UNDERSTAND AND EXPLAIN.

It seems that the residue of dust and debris left over after the formation of the Sun somehow clumped together to form small rocks which coalesced together to form larger rocks and asteroids which, under the influence of gravity (when the clumps achieved sufficient mass to attract further rocks, and perhaps planetoids), to crash into each other, and finally achieve enough mass to become spheres – the most economic geometric shape, (as happens when you blow a soap bubble).

Our early Solar System must have been a very calamitous place with rocks, asteroids and planetoids flying around everywhere and crashing into each other, and the cratering seen on places like the Moon, is evidence of what a dangerous place it must have been. Every planet, moon or asteroid shows evidence of cratering. It seems almost as if someone with a shotgun was regularly blasting into our Solar System to cause this massive cratering.

Even comets, like Halley's show craters on its surface! There are craters, craters and more craters, even craters within craters! Most of this cratering has been lost on Earth due to erosion by wind and water, and just one or two more recent ones, are left. Finally, all this early chaos came to an end and has now become so rare to have almost ceased.

FORMATION OF THE EARTH

It is now thought that in its history, early Earth suffered a collision from another large planet – perhaps the size of Mars that smashed into the Earth at an angle blasting off a large chunk of its surface to form our Moon. Examinations of Moon rocks have concluded that both the Earth and the Moon are made of the same sort of stuff. Early Earth was a great molten ball of fire with meteors and other debris falling on to it. Gradually the heavier elements like iron, sunk towards the centre to form the core and mantle. The surface cooled sufficiently to form the crust. Evidence that the Earth is still hot inside is apparent from volcanoes and hot spots, and even deep mine shafts show a considerable change in temperature compared to the surface. Earth is something like cutting a peach in half where the skin represents the crust, the flesh the mantle and the stone the core – all different in size and makeup.

THE CORE

Our Earth is lucky enough to have a nickel-iron core which spins rapidly in the centre, and works something like an electric generator, creating an enormous magnetic field around the Earth. This shields our planet from energetic particles in the solar wind which would otherwise blow away our atmosphere.

Anyone living in the vicinity of the North Pole can see these energetic particles in the night sky as the Aurora Borealis, or Northern Lights, as they dance around in the night sky in swirls and curtains of glowing light – a pretty and spectacular sight.

THE MANTLE

The Earth's mantle is mainly made up of hot, flowing, plastic rock and makes up most of the body of the Earth. No one has seen the mantle, so we can only guess at its composition and structure. But it is constantly "on the move" and sometimes "hot spots" rise towards the surface, cool down and sink again, something like a lava-lamp. They reckon some of these hot spots made up the volcanos that developed under the Pacific Ocean to form the Hawaiian Islands.

THE CRUST

Above the mantle is, of course the crust, the land masses of which we all live upon. These land masses or continents, are always on the move, albeit at the rate of about half an inch a year. Once upon a time, all the land mass was one vast enormous continent which broke up and began to "drift" in different directions, known as continental drift.

So once the continents of the Americas was attached to Africa and so were India and Australia. India drifted up north and crashed into the land mass of Asia and gave rise to the massive Himalayan range on the borders of the plateau of Tibet. This "drifting" is because the crust of the Earth is divided into something called Tectonic Plates – something like a cracked egg-shell.

These plates gradually move carrying the continents with them, and at times push against other plates, slide or subduct under them. So our Earth, in the dim and distant past, must have looked very different to what it looks like now.

THE OCEANS

Our oceans occupy about two-thirds of our planet. If we were to turn our globe so that the Pacific Ocean is in view, only the land masses of Australia and America are visible at the very edges. Scientists are still at a bit of a loss to explain how all this water got here, as none of our neighbouring planets, or their moons, appear to have any liquid water on their surfaces at all!

Europa, the small moon of Jupiter, may have a liquid water ocean covered by an ice sheet. In addition, recent analysis of the soil of Mars by the robotic vehicle, Curiosity, has revealed that a pinch of the dry soil when heated, yields around 2% of water vapour. This indicates that about a cubic foot of soil would yield about two pints of the precious liquid. It is also believed that deep below Mars's surface lies a layer of permafrost, very much like regions of Siberia on Earth. Recently water is thought to have been detected on the Moon. This water is believed to emanate from deep within the Moon's interior. If these findings are true, it will be a very valuable resource for future astronauts and the setting-up of bases on the Moon and Mars.

Scientists think, perhaps, all this water came from dirty snowball comets that crashed into the Earth in its early history. However, this mechanism is still being debated and researched.

Water may even come from embryo suns. Recent observations of a sun-like star 750 light years away indicate it appears to be shooting jets of water droplets from its poles deep into space, at light speed.

Another possibility is that free hydrogen ions in the solar wind interact with oxygen in the atmosphere and that is why the Earth is so water-rich. It may also explain how noctilucent clouds are formed.

So while, on the surface water may seem very precious and rare, there may be much more of it around than first meets the eye, or the telescope.

THE ATMOSPHERE

Above the surface of the Earth is our atmosphere, or to some who may prefer to call it, the sky. In relation to the size of the Earth, it is very shallow, being only about 75 miles thick. It's like taking a football, dipping it in water, and then comparing the film of water on the surface to our atmosphere. It is thin and very fragile, and it seems we are determined to pollute and destroy this very precious bubble until it is no longer able to support life on this planet. It is fortunate that our Earth has sufficient mass and gravity to hold on to our atmosphere; planets like Mercury have lost theirs long since.

The air in our atmosphere is mainly composed of nitrogen (about 78%) and oxygen (20%). It is oxygen that is vital to life, for without it we could not exist. At the same time, it is a flammable and corrosive gas, causing some metals to oxidise.

It also fans the flames of fires causing them to burn, as every fireman will tell you.

The oxygen in our atmosphere protects us from dangerous ultra-violet light from the sun. It screens most of it out better than any sun-cream, and scatters this radiation to give our sky its pretty blue colour. The atmosphere is denser closer to the Earth's surface, but gradually thins out the higher you go, which is why Everest climbers have to carry a supply of oxygen to enable them to reach the top. Also, the higher you go into the atmosphere, the lower the air-pressure, which is why high-flying passenger planes have to have their cabins pressurised.

When you reach a certain altitude, the atmosphere thins out so much that the sky would appear more black, and even some of the brighter stars would be seen shining in the darkened sky.

[12]

The Inner Planets

THE FOUR INNER PLANETS ARE COMPARATIVELY SMALL, ROCKY BODIES. THE TWO PLANETS BETWEEN THE SUN AND THE EARTH ARE MERCURY AND VENUS.

You will find them in the sky in the east, before sunrise, or in the west, after sunset, depending on their position in their respective orbits around the Sun. The other inner planets are our own Earth and Mars.

THE EARTH

The third planet from the Sun, the Earth is unique as being the only planet - as far as we know - that supports life. Spinning like a top at an angle of 23 and a half degrees and whirling around the Sun, gives us not only night and day, but also our seasons. Our speed in our journey round the Sun is an incredible 67 thousand miles an hour - faster than any spacecraft - completing one orbit in just 365 and a quarter days. At this rate of travel, we could get to Australia from the UK in less than 10 minutes!

Our seasons are caused by the Earth's tilt. When, during the course of its orbit, the Earth's northern hemisphere is inclined towards the Sun, that part of it gets more heat and light, which is our summer. Conversely, when the southern hemisphere is tilted towards the Sun, countries in that part of the world are having their summer. When neither hemisphere is turned towards the Sun (this happens twice in Earth's course round the Sun - the equinoxes) both hemispheres get equal light and heat and experience their respective spring or autumn. Of course the equator - that belt across the Earth's girdle - doesn't have these seasons, because whichever way the planet is tilted, it still gets the same amount of light and heat.

Sometimes this effect can be difficult to understand, but if you have a globe in the house, place it in the centre of a table, and crouch down in front of it till its equator is at eye-level. Now move round the table till the North Pole is pointing towards you. You'll notice that you will be able to see much more of the northern hemisphere in this position. Move a quarter-way round the table again so that you are viewing it side-on. You'll now have an equal view of both the northern and southern hemispheres.

Now move another quarter-way round the table till the South Pole is tilted towards you. You'll see that now you are now viewing much more of the bottom of the globe than the top. In fact you will not be able to see the North Pole at all now, but the South Pole will be fully in view. Finally, move another quarter-way round the table till you are viewing it sideways again. The effect will be much the same as it was before from this position.

So you can immediately realise how important this tilt of 23 and a half degrees is to us living in the northern or southern hemispheres - without it no reasons for seasons!

You'll have seen by your "experiment" that the closer we live towards the equator, the more heat and sunlight we enjoy and, bearing in mind that the Earth is rotating on its axis, (which was not part of the experiment), the days and nights would be of equal length, whatever the season. However, at the Earth's poles it's quite a different story. At the North Pole in summer, light is constantly shining, so it is daylight all the time, whereas at the South Pole, no light is reaching that part of the globe and it's constantly dark. In our winter, the situation is reversed. You'll also notice that in the spring and autumn position, the same amount of light reaches both poles.

The Earth's diameter (cutting it in half, then measuring it across) is just short of eight thousand miles across (7,926 miles to be more precise), but to travel around it, would be almost 25,000 miles (24,902 miles). Travelling around the world non-stop by jet aircraft takes about two days.

If you took a rubber ball and dipped it in water, you could think of the film of water, clinging to the surface of the ball, as our atmosphere. The Earth has an atmosphere made up mostly of nitrogen (over three-quarters), oxygen (under a quarter) and some other gases, which is the air that we breathe. The most crucial is oxygen, for without it, life, as we know it, could not exist on our planet. Our atmosphere doesn't extend very far into space; it is only about 40 miles deep, and starts getting thin and short of oxygen at about seven miles.

Our atmosphere not only protects us from the harmful effects of ultra-violet radiation, shielded out by the ozone layer, but also causes small rocks and objects that hurtle towards us from outer space, to heat and burn up before they reach the Earth's surface. Our atmosphere is a precious and fragile bubble that is constantly under threat by air pollution.

Besides air, the next most precious component for life on Earth is water - our oceans, lakes and rivers. Without water, we could not live either; indeed more than half our bodies are made up of water; without it we would be like dried fish! Most scientists believe that life on Earth began in our oceans of water. Strangely enough, life can exist in water alone. There is no need for any dry land, as any fish will tell you. Imagine a water planet!

THE MOON

The next most familiar thing to the Sun in our Universe is the Moon. Most other planets have moons too, but for a small planet like ours, we have the largest moon by comparison. It is about one-quarter the size of the Earth. It shines with a pale, ghostly glow. Its silvery - and to some, romantic - light is sunlight reflected off it. Although it has the same apparent size as the Sun in our skies, this is caused by the fact that while it is 400 times smaller than the Sun, it is coincidentally 400 times closer. This is very convenient for solar eclipses, as the disk of the moon covers the disk of the sun completely at times, (a total eclipse) and almost completely (an annular eclipse) at others. This is quite a rare event, when the Moon makes its presence known in the daylight sky by coming between our line of vision and the Sun, partially or totally obscuring it.

It is when the Sun is blotted out this way that astronomers can view the Sun's corona, a wispy halo of extremely hot ionised gas that surrounds the Sun and is normally invisible to us because of the Sun's glare. We are lucky that both Sun and Moon have the same apparent size in the sky; otherwise a total eclipse of the Sun would not be possible, and we would miss this rare, and to some, auspicious event.

We also have eclipses of the Moon. This is when the Earth comes between the Sun and the Moon and casts its shadow on the latter. The Moon doesn't disappear completely when this happens, but it is dimmed and sometimes glows a reddish colour.

The Moon circles the Earth roughly once every 28 days, with one side of its face constantly turned towards us. This means it actually turns on its axis once in the same period. To understand this effect, try this simple experiment. Take a teacup, put it on a table, and move it round in a circle. To keep the handle always pointing towards the centre of the circle, you will have to keep twisting the cup. This imitates how the Moon turns on its axis. As a result, we only see one side of the Moon's face. Nobody ever saw the other side of the Moon till the Russian space probe Luna 3 circled the Moon and sent back pictures of it.

The Moon looks different to us at different times of the month by the way it catches light from the Sun, and how we see this effect from Earth. Sometimes it's only a thin crescent, and then night after night, the light creeps across more of the surface until all of it is illuminated. This is the full Moon.

Gradually, the Moon begins to wane, and often you see it very faintly in the blue sky of day, (a gibbous moon), until it fades away altogether. Even when we say there is no moon in the sky, it may actually be there. But with its dark side facing us, it becomes invisible.

The Moon has an important effect on our oceans, seas and other large bodies of water. Because of the Moon's gravitational attraction - in conjunction with the Sun and Earth's own rotation - this causes the oceans to "bulge out" and produces the tides which people living in coastal regions are so familiar with.

The Moon is very much a dead world, with its extremes of temperature between sunlight and shadow and its lack of atmosphere. Its surface is heavily pock-marked by impact craters where asteroids and meteorites crashed into it in the dim and distant past. Since there is no wind, the dust on the surface of the Moon cannot erode them away; hence the footprints of our first astronauts on the Moon will be there for eternity.

Also, since there is no air, it is a very silent place, because sound waves can only travel through air. As you may well know, humans have made six trips to the Moon and landed on its surface. There is an aspiration that if mankind could establish a base there in the future, it will be our stepping-stone to further space exploration and missions to other planets, like Mars.

MERCURY

Mercury is the hardest planet to see with the naked eye. If you're lucky enough you may be able to spot it in the glow of sunrise or sunset. Since it lies between us and the Sun, it is always very close to the Sun in the sky. It's only immediately before sunrise or just after sunset that you may catch a glimpse of it. No wonder it was called Mercury, after the Roman messenger of the gods. It travels fast and is very elusive. On occasion, Mercury actually crosses the face of the Sun and can be seen as a small dark dot, smaller than some of the sunspots. But you must not try to look for this occurrence through binoculars or a telescope if you value your eyesight!

A small rocky planet, Mercury is only 36 million miles from the Sun, and circles it once every 176 Earth days. It spins on its axis very slowly, turning only three times during the course of its journey round the Sun. Being so close to the Sun and turning so slowly, it is being roasted on the side facing the Sun, and the temperature on the equator is hot enough to melt lead (400°C). Meanwhile the dark side of the planet freezes and becomes twice as cold as the Antarctic (-170°C) during its long, long night. A close-up picture of Mercury taken by Mariner 10 in 1974, showed how much like the Moon Mercury is to look at, with a heavily cratered surface. It is only one-and-a-half times bigger than the Moon. There is no chance of any life-as-we-know-it on Mercury.

VENUS

Like Mercury, Venus is another rocky inner planet, but unlike Mercury it is very easy to see in the morning or evening sky, firstly because it is the brightest object after the Moon, and secondly, when visible, it is always found in the eastern or western horizon before sunrise or after sunset. If you're lucky enough to have a telescope in the house, you will see that Venus goes through the same phases as the Moon. It is also the closest planet to us, and was sometimes thought of as Earth's sister planet. Like Mercury, Venus too, transits (crosses the face of) the Sun and this is a rare phenomenon.

The last time these transits took place was in 2004 and 2012, and there won't be another one until the year 2117, a hundred and five years from now. These transits were considered so important that Captain James Cook was sent on his first expedition, on the HMS "Endeavour," to Tahiti with a team of scientists to study this event on June 3rd, 1769.

Venus is completely shrouded in thick cloud, and hitherto no one was able to see its surface. Astronomers thought these clouds could be water-vapour and that Venus could be a water-world. But it turned out that, in fact, these clouds were mainly noxious sulphuric acid, and that this, together with Venus' un-breathable carbon dioxide atmosphere, trapped the incoming heat of the Sun very much like a greenhouse, making the planet's surface intensely hot (475°C) – more than twice as hot as a domestic oven at its hottest!

The Russians found this out when they landed a probe (Venera 4) on the planet's surface on the 18th October 1967. Not surprisingly, the probe was soon frazzled in the intense heat! So, the planet of beauty and tranquillity turned out to be more like something out of hell! Despite several space probes sent to

Venus, its surface was still a mystery, as no telescopes or cameras could peer through the dense cloud. However, the Magellan spacecraft, equipped with its cloud-piercing radar, was sent to orbit Venus in May 1989, and by August 1992, it had mapped most of the planet's surface which has some features similar to Earth, but of course, lacks our oceans, lakes and rivers. As far as life is concerned, scientists agree that it is far too inhospitable a place to nurture any sort of life.

MARS

Mars is also quite an easy object to see when it's in the night sky. It usually gets quite bright when it's close to us, and shines with a reddish glow. Although it is classed as one of the inner planets, it is not between us and the Sun like the other two, but lies between us and the outer planets of the Solar System. Like them, it travels across the ecliptic from east to west. Mars circles the Sun once in nearly two Earth years (687 days) and it has two tiny, irregular shaped moons called Phobos and Deimos.

Mars is quite a barren world, with cold airless deserts, mountains and extinct volcanoes, one of which, Olympus Mons, is far larger than anything we have on Earth. Mars is probably the most studied of all planets because its thin carbon dioxide atmosphere is almost devoid of clouds. When there are clouds, these are usually clouds of dust because Mars has no apparent water. It has polar ice caps but these too are frozen carbon dioxide. Satellites like Mariner were sent by us to photograph it, and two spacecraft, Viking 1 and Viking 2, actually landed on the surface in July and September 1976 respectively, to take photographs at ground level.

It showed a red desert with scattered rocks, and a pink sky. In 1997 the spacecraft Pathfinder bounce-landed on the surface, and a small robotic vehicle was sent out from the lander to investigate nearby rocks. In 2004 two robotic rovers, "Spirit" and "Opportunity" landed on the surface of the planet and began explorations. More recently, in 2012, another robotic vehicle, "Curiosity," landed on Mars and is conducting various scientific experiments besides sending back pictures.

At one time Mars was thought to be inhabited by Martians, because in the early 1900's an astronomer called Percival Lowell popularised a theory that some peculiar markings he noticed on the face of the planet were artificial canals for irrigation. These were never verified and later observations have not shown any evidence of this. In 1996 a meteor ALH84001, thought to have originated from Mars and found in Antarctica, where it had lain on the frozen surface for 13,000 years, was, on microscopic examination, believed to contain signs of tiny fossils. This caused great interest and is still under investigation.

It might prove that Mars may once have contained water, and that life, however primitive, could once have existed on the planet. Besides Earth, Mars is the most explored planet in the Solar System, and it's hoped that further NASA and other explorations will give us a better understanding of Mars, and the secrets it is still hiding. An orbiter is already in place around the planet to map its entire surface.

PHOBOS AND DEIMOS

Unlike our own Moon, Mars' satellites are small, lumpy, potato-like objects more akin to asteroids than true moons, measuring only a few miles across. They were probably asteroids that were captured by the gravity of Mars at some point in time.

[13]

The Outer Planets

THERE IS A LARGE GAP BETWEEN THE INNER AND OUTER PLANETS, AND THEN, EXCEPT FOR A BELT OF ASTEROIDS – SMALL AND LARGE LUMPS OF ROCK - COME FOUR ENORMOUS GAS GIANTS - JUPITER, SATURN, URANUS AND NEPTUNE.

The fifth outer "planet" is Pluto, which is a tiny, frozen, rocky oddity on the very fringe of the Solar System and, after some debate, Pluto was re-classified as a "dwarf planet" in 2006.

JUPITER

Jupiter, is almost unmistakable - a bright shining lamp in the night sky. Through a small telescope, the planet will look quite round, and you will be able to see, just like Galileo did, Jupiter's four moons - tiny little pinpoints of light, every night in a different position as they dance around the planet. Sometimes all four are in a straight line; at other times you may see two on one side of Jupiter and two on the other, or you may only notice three moons because one is hidden behind Jupiter, or is passing in front of its face.

No wonder Galileo got his idea of how the Solar System worked by observing Jupiter and its moons, because it behaves like a Solar System in miniature. Jupiter has a beautifully marbled appearance, with a great red spot - in reality a gigantic storm - that moves across its surface. Some of these features can be seen through a small telescope, but it was the Pioneer and Voyager fly-bys that first revealed how beautiful the planet is.

Jupiter is over five times our distance from the Sun, and it takes nearly twelve of our Earth years to complete one circuit. It is the first of the great "gas giants" of our Solar System, because unlike the planets we have talked about so far, these do not have a solid surface; they are great balls of mostly hydrogen and helium gas that becomes denser and denser as you go deeper and deeper into the planet, until under the enormous pressures, the gas becomes liquid and eventually the liquid is thought to become metallic, unlike any substance we know on Earth. In 1996, a probe was actually sent into the Jovian (Jupiter's) atmosphere. Its parachute opened and signals about the environment were sent back to Earth before it was eventually crushed out of existence by the enormous pressures it encountered.

Jupiter is the largest planet in the Solar System. It could easily swallow 1,800 Earth's, and contains as much matter as all the rest of all the planets put together! Jupiter's huge gravity acted like an enormous vacuum cleaner in the Solar System's early history, sweeping up most of the cosmic debris - the remnants of matter left over from the making up of the Solar System - with its enormous gravitational tug.

Were it not for this, our planet and others could have been battered out of existence by all this debris.

Although Jupiter is composed of gas, the moons around Jupiter are rocky like our own Earth. There are several of these moons, but the largest and most well-known are the Galilean moons, Callisto, Ganymede, Europa and Io, which we shall talk about separately. There is much interest in these moons, because unlike Jupiter itself, it would be feasible in the future to land craft on their surfaces, in preparation for a possible robotic exploration. There are also suggestions that at least one of these moons may harbour life.

CALLISTO

The outermost of Jupiter's moons is a bit larger than our own moon. It has a dark surface which is very heavily cratered, in fact it is the most cratered object so far seen in the Solar System.

GANYMEDE

The next furthest from Jupiter it is the largest moon in the Solar System, in fact it is larger than the planet Mercury. Ganymede too has a heavily cratered surface, but there are also bands of parallel grooves that extend for thousands of miles and sometimes intertwine with each other.

EUROPA

The third moon in towards Jupiter, Europa is the smallest of the Jovian moons and unlike the others, is as smooth as a billiard ball. It has a bright surface criss-crossed with darker lines, which seems to suggest a sea frozen over with cracked ice. If there is indeed water under the ice, there is a strong possibility that there may be some form of primitive life there too.

IO

The most geologically active body in the whole Solar System besides the Sun! Io's surface looks like a pizza with patches of yellow, orange and red. Its volcanoes spew gas and dust high into the atmosphere, and although surface temperature of this moon is freezing according to Earth standards, near these volcanoes the temperature may be in the balmy 80's. Poor Io is very close to Jupiter, and the tidal forces exerted by its huge mother are so great, that its interior is being pulled to and fro, causing a lot of frictional heat. This is what causes all that volcanic activity.

SATURN

Saturn, the next great gas giant, is second in size only to Jupiter and shines a pale gold. Saturn is about nine-and-a half times further away from the Sun than we are, and it takes over 29 years to make one orbit.

Close-up pictures of Saturn taken by the Pioneer and Voyager spacecraft showed practically no surface features on this golden planet, and special techniques had to be used to enhance whatever features could be distinguished. But what makes Saturn the most beautiful of all the planets is that it is surrounded by spectacular rings that are made up of millions and millions of small rock-like and icy particles that revolve around the giant planet.

Again, if you are lucky enough to have a look at Saturn through a small telescope, you will see these rings. How well you see them depends on the tilt of the planet when viewed from Earth. In 1951 for example the rings were edge-on to us, and the fine line invisible. Although these rings are many thousands of miles wide, they are only a mile or so thick. In 1960 the southern pole of Saturn was turned towards us so we viewed the rings from underneath. In 1967 we were back to the thin line, and in 1975 we were looking at them from above. In a telescope you may also catch a glimpse of Titan, the largest of the Saturnian moons, and maybe some others as well, for Saturn has many smaller moons and satellites.

SATURN'S MOONS

Titan is a large moon, second only in size to Jupiter's Ganymede, and an interesting one because it seems to have a substantial atmosphere. It is possible that life could exist there. In 1997 the Cassini spacecraft was dispatched on a seven year journey to Saturn.

On board Cassini was Huygens, a space probe which soft-landed on the surface of Titan. It was equipped with cameras and scientific instruments, and took photographs through the smoggy atmosphere to reveal a rather "earth-like" surface, with seas of probably liquid methane. This is the first time one of our space probes has landed on an alien moon. The larger of the other moons are Mimas, Enceladus, Tethys, Dione and Rhea. Mimas is interesting, because towards its North Pole is an enormous crater, which scientists think was caused by a huge impact of an asteroid.

URANUS

Lying 19 times the Earth-to-Sun distance away is another gas giant, Uranus, the seventh planet from the Sun. It is a strange planet - strange because it is the only one that is tilted in such a way that each of its poles in turn is pointing head-on towards the Sun; this means that if you were able to live on the planet then you would have 21 years of "day" and 21 years of "night."

It also takes 84 Earth years to complete one orbit around the Sun. Like Saturn, Uranus too has rings, but they are very narrow and were only seen when Voyager 2 sent back pictures.

The rings, incidentally, are face-on towards the Sun. Uranus is so far away from us and is such a small object amongst the myriads of stars, that its greenish disc can only be picked out with a telescope, and even then it is difficult to find unless you know exactly where to look for it. It was not until 1781 that it was discovered by Sir William Herschel. Like the other gas giants, Uranus has its complement of satellites.

SATELLITES OF URANUS

Titania is Uranus' largest moon. It was also discovered by William Herschel in 1787. It's an icy, cratered body. A geologically exciting moon is Miranda. This moon is smaller than the others and the closest to the planet, and has a bizarre surface. It is thought that Miranda was once broken up by another astronomical object, and then melded together again.

It is also thought that something very big, in the dim and distant past, disturbed Miranda's mother planet, Uranus, to place it at this peculiar angle. Among the other moons of Uranus are Ariel, Umbriel, and Oberon.

NEPTUNE

About the same size as Uranus, and a pretty blue colour, Neptune also appears so small in the night sky, that it was not discovered visually, but rather it was predicted by astronomers that there must be an eighth planet to account for a slight perturbation in the orbit of Uranus. They reckoned something big out there was having a gravitational pull on Uranus. So when they began searching, an eighth planet was found in 1846 and it was named Neptune, after the Roman sea god.

The same year, it was also discovered that Neptune had a satellite, slightly smaller than our Moon, and this was named Triton. Because even the most powerful telescopes on Earth were not able to get "close enough," no one knew much about the surface features of Neptune till Voyager 2 on its journey, passed close to the planet in 1989, and sent back pictures.

Scientists were quite amazed to see white clouds scudding across the face of the blue planet. Previously they thought Neptune was too far away from the Sun to have a weather system.

Neptune is over 30 times as far away from the Sun as we are, and it takes over 164 years to complete one orbit.

NEPTUNE'S SATELLITES

Among Neptune's many moons, Triton is Neptune's largest. It was discovered less than a month after its mother planet in 1846. It is slightly smaller than our own Moon, and has a cratered and cracked surface, almost like a jumble of ice and rock.

When Voyager 2 passed close to the moon and sent back pictures, they revealed active geyser-like eruptions spewing nitrogen gas and dust several miles into space.

[14]

Flotsam and Jetsam

THERE ARE SEVERAL REGIONS IN THE SOLAR SYSTEM THAT CONTAIN WE CAN THINK OF AS 'LEFT OVER BITS' THAT NEVER QUITE MADE IT INTO PLANETS.

ASTEROIDS AND METEORS

In the large gap between Mars and Jupiter is a band of asteroids - lumps of rock really - some large, some small, all swirling round the Sun in a swarm. Some of the larger asteroids, like Ceres which is nearly 500 miles long, can be seen by powerful telescopes. One of these asteroids, Gaspra, an irregular 12 mile long piece of rock, was actually photographed at close range by the space probe, Galileo in 1991 on its journey to Mars.

Besides the asteroids and meteors in the belt, there are others that wander around the orbits of Mars and our own Earth, and some have impacted with Earth in the dim and distant past. The most well-known example is the Barringer Meteor Crater in Arizona.

PLUTO

The ninth and last planet (which is now thought of as a dwarf planet or planemo) of the Solar System, Pluto is paradoxically the smallest of all, smaller even than our Moon! It is not a gas planet either, but a rocky one. It was not discovered till 1930 and was named after the brother of mythological Zeus and Poseidon, the Greek counterparts for Jupiter and Neptune, and not after the Disney dog! It is a strange little icy world, on the very fringe of the Solar System, 39 times further away from the Sun than we are, but it has an eccentric orbit which sometimes brings it inside the orbit of Neptune.

Pluto takes over 248 years to complete one orbit of the Sun. It has a moon called Charon which was discovered in 1978 and the two rotate around each other. This is because Pluto has such a small gravitational pull on its comparatively large moon that the two are attracted to each other and spin round a common orbital point.

COMETS

Comets are occasional visitors to the inner Solar System. They are small celestial bodies of frozen gases, grit and dirt in what has been termed "a dirty snowball," and occasionally decide to visit the neighbourhood of our Sun.

It's thought that they live in two regions, the Kuiper Belt and the Oort Cloud. The Kuiper Belt is a relatively flat region outside the orbit of Neptune which is the home to short period comets that return every 20 to 200 years or so. The Kuiper Belt extends from around 30 to 50 AU.

The Oort Cloud however is a spherical shell of comets which lies far, far beyond the orbit of the outermost planet. It is thought that it lies around 50,000 AU away. This is nearly a light year away and a quarter of the distance to our nearest stellar neighbour, Proxima Centauri.

From their home in the Kuiper Belt, it's reckoned that some get nudged out of place by perturbations inside our Solar System and begin a long, long journey towards the Sun. These journeys can take thousands of years. Comets from the Oort Cloud may be disturbed by the motions of other solar systems and might have an orbital period of thousands or even millions of years.

Irrespective of their source, as comets become visible to astronomers (generally amateur comet-hunters), comets are usually named after the astronomers who discovered them in their telescopes. When these comets come nearer the Sun, they warm up and sometimes develop very long tails of dust, gas and other cometary material. These tails, do not plume out behind the comet, as is popularly thought; instead they are blown out by the "solar wind" our Sun generates, and so always point away from the direction of the Sun.

These visitors put on a great show in the night sky with their streaming tails, and to see a really bright one is a fine and rare sight. Some comets have very elongated orbits, and after they have swung round the Sun, they disappear from view, never to return, or if they do, thousands of years later. Some come back quite regularly though, and the most famous of these is Halley's Comet, named after its discoverer Edmond

Halley, a contemporary of Sir Isaac Newton. He found that the orbits of "some" comets observed in 1531, 1607 and 1682 was really one and the same comet, and so he accurately predicted its return in 1759. Halley's Comet is therefore a regular visitor every 76 years, and is the 'once in a lifetime comet', because it usually appears once in a person's life span. It has a comparatively short orbit which makes these 76 year visits possible.

On its last visit in 1986, a spacecraft Giotto was despatched to get a closer look at it. Before it was battered with particles from the comet's tail, it sent back pictures of something that looked like a large peanut, surrounded by a halo of glowing gas. One could even spot a crater on that peanut! However, on the last return of Halley's Comet, many observers were disappointed because it was not well placed for viewing that time round. On the other hand, comet Hale-Bopp, like its famous predecessor comet Hyakutake, put on a spectacular show in the spring of 1997, trailing a magnificent tail. It was a large and active comet too, and a joy for many observers to see this solar visitor grace the night sky with its beauty and brilliance.

In ancient times, comets were thought to be harbingers of doom, and this is understandable, even today, because they have such erratic habits. One day, a comet may crash into our little planet. Some scientists say this already may have happened in the distant past, which is why the dinosaurs disappeared so quickly in their hey-day.

A comet crashing into a planet was witnessed in 1994 when Shoemaker-Levy 9 broke up before plunging into Jupiter, a rare and spectacular sight, made possible by the excellent observations through the Hubble Space Telescope.

Now the Solar System eagerly awaits the arrival of a new comet in our skies in November 2013, and that is Comet ISON. It is hoped that this new visitor will be bright and dazzling with a long tail, but we won't know until it gets to around the orbit Mars and beyond a place known as the "Frost Line" where the nucleus will start to melt. We hope it won't disappoint us as sometimes comets do, as many promise to be bright and prominent in the skies but fizzling out at the last moment. After the comet makes its pass towards the Sun in November 2013, the Earth will also pass through its tail in January 2014 producing some amazing meteor showers.

[15]

Beyond the Solar System

WHEN WE EXAMINE THE SCALE OF THE SOLAR SYSTEM, IT IS DIFFICULT ENOUGH FOR US TO IMAGINE THE LARGE DISTANCES BETWEEN THE SUN AND PLANETS. WHEN WE GO BEYOND THE SOLAR SYSTEM, THE DISTANCES BECOME ALMOST INCONCEIVABLE.

Our own Sun has several stellar neighbours that lie comparatively close to it. The nearest of these is Alpha Centauri which lies in the constellation of Centaurus, the Centaur, and can only be seen from southern latitudes. To the naked eye, it shines as a single bright star, but telescopic observation reveals that it is really two stars, Alpha and Beta Centauri, orbiting around each other. The brighter one, Alpha Centauri, is very similar in nature to our own Sun. There is also another sun, Proxima Centauri, a red dwarf that orbits around the other two and this star is presently the closest to our own Sun. The distance to Proxima Centauri is calculated as a little under 4¼ light years away, that is to say that it takes light, travelling at the speed of over 186 thousand miles per second, (186,282 to be precise), that amount of time to reach there.

So, if it were ever possible for us to build a spacecraft which could travel at about that speed, that is how long it would take to reach our neighbouring star.

At the moment, with our present rocket technology, it would take at least 60 thousand years for even our fastest spacecraft to reach the vicinity of Proxima Centauri. When you consider that the early history of our civilisation – like the building of the Pyramids – only goes back about 5 thousand years, it will give you some idea of just how far away our nearest star is.

There are only about 20 other stars that lie less than about 12½ light years away from us, and a couple of the more well-known ones are the two "dogs" following on the heels of Orion. These are Sirius and Procyon, and they are amongst the brightest in the night sky. Sadly, this is not true for some of the other nearest stars, for "near they might be" but "bright they are not," like Proxima Centauri which is a dim little red dwarf, invisible to the naked eye.

Conversely, most of the brightest stars we can see in the night sky are vast distances away from us. For example Vega (in Lyra) is 26 light years away; Antares (in Scorpio) 420 light years away; Rigel and Betelgeuse (both in Orion) 600 light years away; Deneb (in Cygnus) an incredible 1,400 light years from us. Even the four stars that make up the pan of the Big Dipper (Ursa Major – or call it the Great Bear if you like) are 62 to 75 light years away. This is because they are extremely bright, some of the more distant ones being in the supergiant class.

It's like a friend standing 20 yards away from you on a dark night holding out a candle, and another friend, a mile away, with a great big searchlight. There are no prizes for guessing whose light will appear brightest!

So, how do we cope with these unimaginable distances when it comes to inter-stellar space travel? We have become so used to seeing science fiction movies and series on television, where the space travellers flit from star to star in a trice, that we have almost become blind to the realities of space travel. Faced with the immense technical problems this presents, one quickly comes down to earth with a bump!

MEASURING DISTANCES IN SPACE

With these vast distances in space, you might well ask, how do astronomers measure these huge distances? Well, to start with, the nearer stars are not such a problem. They measure them by their parallax. It is the same effect when you hold a finger at arm's length, and look at it through one eye, then the other, and notice how much it has shifted against the background.

For nearby stars, this method works by taking one measurement when the Earth is in a particular part of its orbit, and another 6 months later when it is the opposite side, then calculating the star's parallax – how far it has moved against the background of other stars By measuring the angle of shift and using a trigonometrical calculation they work out the star's distance in parsecs (one parsec = 3.262 light years) which they find more convenient than working in distances of light years.

For more distant stars the astronomers compare the actual brightness of a star (how bright it really is) against it apparent brightness (how bright it appears to us), and knowing how light falls off with distance, as candlelight does the further it is away from you, can measure how far away it is. Some stars, called Cepheid variables, fluctuate in brightness periodically in a regular way. Measuring this change in relation to period and brightness, they use these stars as "cosmic candles" to work out some stellar distances and determine cosmic distance scales.

For even more distant objects, they measure a star or object's "red shift" on a spectrogram by looking at the light of its spectrum. This is easy enough to do yourself with a bright object like the sun, when a prism will split up its light into a rainbow of colours. Well, on a spectrogram, small dark lines, called "spectral lines" can be seen lying across the colours, and these represent the chemical elements of the star. In the normal state of affairs, these stick to certain regions of the spectrum. But when light from very distant objects is observed, these "spectral lines" are shifted towards the red portion of the spectrum. They call this the star or object's "red shift," and by measuring how far the spectral lines have shifted towards the red from their normal places, they can tell how far away the star or object is away from us, and indeed for very distant galaxies and quasars, how quickly they are moving away from us in an expanding Universe.

When light moves towards you, it appears "bluer" and when it moves away from you "redder." This effect is only measurable in scientific experiments, but it is easier to distinguish when it comes to sound.

For example a train whistle or car horn will sound higher in pitch when it is travelling towards you; then as it passes, the pitch will become lower. This is known as the Doppler Effect.

[16]

Reaching for the Stars

WE ARE IN FACT ALREADY SPACE TOURISTS, AS WE TRAVEL THROUGH THE COSMOS ON 'SPACESHIP EARTH' AT THE INCREDIBLE SPEED OF AROUND 67 THOUSAND MILES PER HOUR ON OUR ANNUAL TRIP AROUND THE SUN.

Meanwhile, the Sun is pulling it and all our planets around the Milky Way at a speed of about 485 thousand miles per hour! So we're really travelling! Therefore, the word "motionless" really means "still in relation to something else" because everything is on the move! So the cup of coffee on your desk is motionless in relation to the room; the coffee cup on your table in a jet aircraft is motionless in relation to the plane, but not to an outside observer, because it is really jetting through the sky at about 500 miles per hour!

Travelling through outer-space, away from our safe haven of the Earth, has intrinsic problems for us humans. We are simply not designed for space travel. First zero-gravity, as every astronaut will tell you, has a detrimental effect on the muscles and bones of our bodies.

Next, harmful radiation from the Sun is not screened out by our Earth's atmosphere, and will damage the cells in our bodies. Then there is, of course, no air in space, so we have to take our own breathable air with us in the spacecraft or spacesuits. Also, there is no atmospheric pressure as is found on Earth, so without the protection of a pressure-suit or pressurised cabin, we would not survive for long, though perhaps not blow up in a rather messy explosion, as sci-fi writers love to depict!

With our technology, we have managed to overcome these problems for short-haul space flight or living for about a year or so in an orbiting space station like Mir and the International Space Station (ISS). However, travelling to our nearest neighbour Mars, would take about 9 months, and that's only to get there, so it's tough asking our would-be astronauts to endure these conditions for as long as a three-year round trip! Not only would it place a great strain on their bodies, but there are also the psychological aspects to consider.

The only answer to the problems of inter-planetary space travel is to get there faster – say in days rather than months. This is not impossible, as Captain Cook would have thought if you had told him that his three-year journey to Australia could be done within a day if he could wait a couple of hundred years. But to achieve this, much more powerful engines would be required with perhaps a new energy source, as our present rocket technology is simply not efficient enough. Also living space and conditions would have to be made more adequate and comfortable for our travellers.

Again imagine what Captain Cook would have thought about a journey to Australia on a modern ocean liner instead of the confines of his tiny sailing ship, Endeavour.

A popular alternative concept is to put our would-be astronauts into deep hibernation, in special capsules with life-support systems, and this theme has been used in many sci-fi films. Certainly this concept would be appropriate for voyages to our distant planets where the journey would only take years, and not hundreds of thousands of years as in the case of a star.

For inter-stellar (between the stars) voyages, we have to go into the realms of science fiction. Supposing it were possible for us, sometime in the far, far future, to capture a fairly large asteroid – say about a mile or so long, and tether it in Earth orbit, then build a mini city on it, enclosed in domes of transparent material specially treated with something to protect us from the harmful radiation in outer-space, with factories, farms, power plants – even parks and tree-lined avenues, so that we could comfortably live on it for many, many generations. We would have to invent some sort-of space elevator to carry all the materials required for this massive building project in space.

On this contrived mini-world some sort of artificial gravity would have to be developed to make life comfortable. Very powerful engines would have to power this craft through space. Even if we could power it enough to reach a speed of a comet, say 158 thousand miles per hour, it would be at least three times the speed we can presently achieve with our present rocket technology.

When it got to its destination – the planet of a not too distant star, it would have to go into orbit around it, and using space-shuttles on board, travel to the surface of the planet to explore it.

But you may well ask "Why would we want to travel to another star anyway?" The only feasible answer would be if our planet, or indeed the Solar System, was under threat, and we would be forced to migrate to another world in order to survive. To travel all this distance to merely satisfy our curiosity and to see if we have any stellar neighbours, who might be hostile to us anyway, is not a sufficient reason!

If the voyage should take several thousands of years of many, many generations of the human travellers. It may come as a surprise if our intrepid pioneers were to find human beings already living on the planet, and waiting there to greet them, having arrived from Earth at near light speed in the intervening years, using the advanced technology for inter-stellar travel they might have developed since their forefathers began their long, long journey!

Meanwhile our two little space-craft Voyager 1 and Voyager 2 are continuing their journey through outer space, having completed their tasks of taking close-up photographs of our outer planets, Jupiter, Saturn, Uranus and Neptune which had never been seen in such detail before. Launched in 1977 they are now the most distant spacecraft that have ever travelled, 125 and 101 AU from the Sun, respectively. Presently, they are still able to receive commands from Earth and transmit data back to the Deep Space Network.

In time to come, they will eventually reach the boundary of the Solar System, the Oort cloud, which lies about a light year out from the Sun, and is a sort-of shell of comets and other debris left over from the formation of the Solar System.

Who knows, if they continue to travel undisturbed for hundreds of thousands of years, someday they may reach the vicinity of another star and be captured by intelligent beings. If so, both these craft have twelve-inch gold-plated copper disks attached to them that contain sounds and images that were specially selected to portray the diversity of life and culture here on Earth. However, it is more likely that they will be captured by the gravity of the alien star or simply burned up if they ventured too close.

Our Pioneer 10 and 11 spacecraft too, launched before the Voyagers but not so far distant, since they are travelling more slowly, will also find their way into the heliosphere (the limit of the Sun's influence) and venture beyond into deep space. Like the Voyagers, they too flew past Saturn and Jupiter and sent back the first images, though not as detailed as Voyagers'.

Both Pioneer 10 and 11 are carrying golden plaques, depicting a man and a woman and information about where they originated from, so in the event that they were ever captured by extraterrestrials, they would become aware of us. Whether this was a wise decision is a moot point!

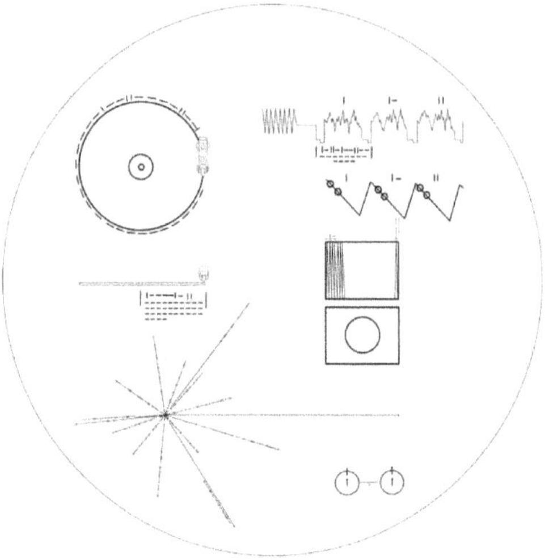

The golden disk attached to Voyager 1 and 2 spacecraft containing scenes, greetings, music, and sounds from Earth.

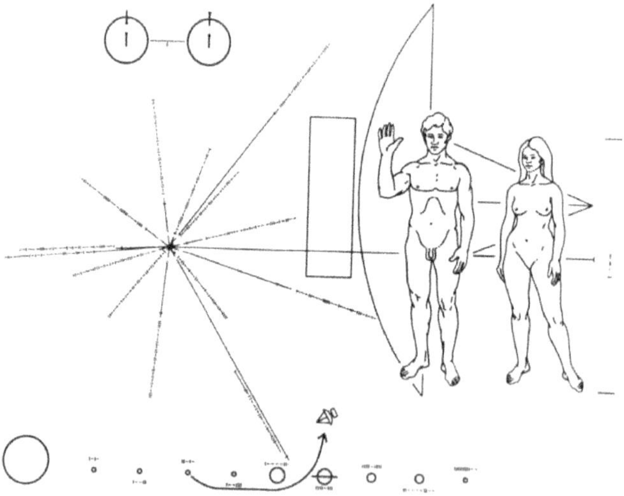

The Pioneer 10 & 11 plaque, depicting the naked human forms of man and woman, plus other signs and symbols and a crude map of our Sun and planets, showing The Voyager spacecraft leaving planet Earth.

[17]

The Milky Way

OUR SUN, EARTH, PLANETS AND ALL THE OTHER STARS LIVE IN A PART OF THE UNIVERSE CALLED THE MILKY WAY, A SORT OF 'ISLAND OF STARS' IN SPACE.

It got its name because on a very clear night, far from the glow of city lights, if you look up at the sky you will see a sort of hazy band of light that looks a little like a milky river of cloud. This is our edge-on view of our Galaxy, and the hazy band of light is in fact, millions upon millions of stars. If we were able to view our Galaxy from above, it would look like a beautiful Catherine-wheel, (like our nearest galaxy Andromeda) with spiralling arms that seem to want to wind up in the centre. But from our viewpoint, somewhere towards the edge of one of these spiral arms, we are looking in towards the galactic centre in the constellation of Sagittarius, and all the stars in the plane of our vision look concentrated and cluttered-up.

If we turn around and look in the opposite direction from Sagittarius, then we are looking away from the centre of our Galaxy, and we see the stars of the other spiral arms beyond ours.

Of course, we see stars all over the sky, and indeed every pinpoint of light is a star (except for the four more easily visible planets when they are in the sky), but they are concentrated in the hazy band of the Milky Way. If you see a photograph taken through a powerful telescope aimed at a tiny region of the Milky Way, you would hardly see any "blackness" of space; instead there would be millions and millions of tiny speckles of starlight covering the entire frame.

The Milky Way is a "barred spiral" type galaxy, and our Sun lives in one of the spiral arms, the Orion arm, made up of millions of stars and nebulae. They wind outwards from the central bulge of our Galaxy's centre that would appear something like a humble fried egg. We live about a third of the way out from the centre, where a supermassive black hole is thought to exist, causing all the stars in the Galaxy to swirl around it. Our Sun makes its journey, with all the other stars, completing one orbit of the Galaxy in about 226 million years. This is known as a "galactic year." Since it was formed, our Sun has circled the Milky Way about 20 times. All the stars we see in the night sky belong to the Milky Way, also gas, dust, nebulae and bright globular clusters, composed of millions of stars in gigantic spheres, which hover above the plane of the Galaxy. Also hovering above the Galactic plane, are our two companion galaxies, the Magellanic Clouds.

It is impossible to count all the stars in the Milky Way, but it is estimated that there are at least 200 billion (200 thousand million) of them. Now, this is a very large number, and like all large numbers, tends not to mean very much to the average person.

The best way to understand what a figure like this means is to imagine you could share out all the stars in the Milky Way with every man, woman and child in the world. You could afford to be generous and give each at least 28 stars, and you would still have plenty left over.

Every star in the Milky Way is a sun that shines with its own light. Our Sun is just one star amongst all the rest, and a rather small star at that. Our place in the Milky Way is about a third of the way out from the centre where, lost amongst the myriads of stars, our little Sun sheds its pinpoint of light.

Of course to accommodate all these billions of stars, the Milky Way has to be very large. To deal with these very large distances, astronomers have devised a measurement called the "light year." It simply means the distance which light, travelling at the incredible speed of 186 thousand miles per second, will cover in the period of one year.

It is estimated that it will take 100 thousand years, travelling at the speed of light, to cross from one end of the Milky Way to the other. Our Sun is reckoned to be about 30 thousand light years away from the Galaxy's centre. As far as thickness is concerned, Astronomers estimate that the Galaxy is 2 thousand light years thick.

Besides its millions of stars, the Milky Way is also made up of an abundance of gas and cosmic dust known as nebulae. These misty patches of light can sometimes be discerned with the naked eye, but are best seen through binoculars or a small telescope.

There are many beautiful pictures in astronomical magazines of these nebulae, in glowing shades of blue and red and green. However, not all these misty patches are nebulae; many are other galaxies

STARS OF THE MILKY WAY

Before we go any further, we'll have to look at a star. To see the best example of a star, simply look at our familiar Sun, preferably when it is just rising or setting. It is our brightest and nearest star. It is by studying the Sun that scientists have worked out what makes stars tick.

The Sun is our most studied star and in recent years it is under constant surveillance. Two dedicated satellites, Stereo A and Stereo B were launched in 2006 and placed in different orbits round the Sun, outside Earth's orbit. This is to enable them to take stereo images of the Sun and monitor its behaviour, to warn us of any large incoming solar flares and coronal mass ejections, that could cause power surges on Earth and disrupt our electric supplies, and even frazzle the electronics of our precious satellites that we have become so dependent upon for things like satellite TV signals and our GPS networks.

BINARY AND MULTIPLE STAR SYSTEMS

We are normally more familiar with the concept of a single star, perhaps with planets orbiting around it as in our own Solar System. But in reality, it is quite rare to find a single sun, and more so to find one with planets! In fact the reverse is true. Most stars nearer the Sun, and those that can be comfortably studied, are binary or multiple star systems.

Some are "close" binaries (i.e. their orbits are close to a central point), and some "distant" binaries, where they are further apart. It became great fun for amateur astronomers to look at these stars with their telescopes to see whether they could split the very close binaries, as it requires more and more powerful instruments to do so. The star Sirius (the Dog Star), is a close binary – too close for splitting with amateur telescopes; whereas the famous "Double-Double" (Epsilon Lyrae) in the constellation of Lyra (The Lyre) is a remarkable multiple star.

Through even a small telescope the two stars show up quite clearly. It takes a more powerful instrument to discover that in fact each of these two stars is a close double itself! Binary and multiple star systems could create a problem for planet formation because of the turbulence of the gravitational forces exerted on each other. Our nearest star, Alpha Centauri, is a multiple star system of three stars, so although one of the suns is very similar to our own, chances are that planet formation would not be very probable.

GLOBULAR CLUSTERS

These are great big clusters of stars – hundreds of thousands of them – all densely packed together in a great big ball, usually found hanging above the plane of the Milky Way. There are about 150 of these Globular Clusters attached to our own galaxy, and they have been seen in other galaxies. They are thought to be very old stars formed at about the same time when our galaxy was very young. Since they are so densely packed the distance separating the individual stars would not be very great.

OPEN CLUSTERS

The stars in open star clusters are more spread out than in a globular cluster. An open star cluster may contain any number of stars from a few dozen to several thousand, and the majority will be hot, young stars. A very well-known open cluster is that of the Pleiades (M45) or "Seven Sisters," which lies in the constellation of Taurus (the Bull). Seven of the brightest stars are visible to the naked eye, but binoculars or a small telescope will reveal dozens more. These stars are so young that faint nebulosity, from the gas and dust they were formed from, still hangs around them in a haze. Other open clusters like the Hyades, also in Taurus, and the Double Cluster in Perseus are a superb sight in binoculars for any aspiring astronomer.

THE MAGELLANIC CLOUDS

The two Magellanic Clouds – the Large (LMC) and the Small (SMC), can only be observed from the Southern Hemisphere, where they are conspicuous in the night sky. They are in reality, satellites of our own Milky Way, and are irregular in shape. The LMC lies about 160 thousand light years away and the SMC is around two hundred thousand. Observing with binoculars, masses of stars can be seen, and the Tarantula Nebula lies within the LMC, where swirling clouds of gas make it a must for all star gazers.

NEBULAE

These are enormous clouds of gas and dust within the Milky Way, and are regions where star formation occurs. The gas and dust clump together to form larger and larger particles, eventually growing to a size big enough to form stars.

There are both bright and dark nebulae – bright when lit up by embedded stars, and dark when appearing in silhouette against brighter nebulae. The Horsehead Nebula in Orion is a good example of dark nebula; the great Orion Nebula a good example of a bright nebula. Both are wonderful objects to be observed with a telescope, and pictures taken by the Hubble Space Telescope reveal them in all their glory.

[18]

The Life and Death of Stars

LIKE ALL LIVING THINGS, STARS GO THROUGH A PROCESS OF BIRTH, LIFE AND DEATH.

They form from clouds of hydrogen gas and dust. This gas and dust is very low density stuff, only containing a few atoms in a shoe-box full of space, and they tend to hang together in clouds called nebulae. Sometimes, these nebulae start to coalesce, under the influence of gravity, and wind up as a gigantic swirling mass of matter.

As more and more matter condenses, the core of this gas-ball begins to warm up and will form into something scientists call a protostar (a star embryo), and begins to radiate energy. But at this stage, there is a continuous battle against the forces of pressure and gravity - the inclination of gravity to squeeze the protostar down and the internal pressure within the protostar to keep it up. At the same time too, radiation emitted by the protostar in the form of light and heat, drains it of energy. This is something like trying to inflate a leaky balloon.

You puff to fill it with air, but the elastic force of the rubber keeps trying to contract it. Gravity is winning at the moment and the protostar continues to contract. However to tip the scales, very, very high temperatures are beginning to build up in the protostar's core, and when enough heat is finally reached, nuclear reactions begin - you now have your cosmic bicycle pump! Hydrogen, of which the protostar is made up, is converted to helium, in a process called nuclear fusion, giving off an enormous amount of energy in the process. This hugely helps the protostar win its fight against collapse. Its internal pressure is now equal to the effect of gravity, it is now in balance, and shining brilliantly, becomes a true star.

How long a star lives and with how much energy it will use depends very much on the amount of matter it has been able to accumulate - its mass. Hot, young stars, much more massive than our Sun, tend to burn very energetically and use up most of their fuel quickly, whereas less massive stars burn their fuel more steadily and live far longer. Our Sun is expected to "live" for another 5,000 million years before it starts its downward spiral towards a slow and comparatively gentle death.

SOME TYPES OF STARS

Stars come in all sizes, and astronomers grade them, as a jeweller would grade gemstones, for their size, colour and brilliance. They use a graph called the Hertzsprung-Russell diagram where the actual brightness of stars is plotted against their temperature.

G-TYPE STARS

Quite rare in the Milky Way are G-type stars like our own Sun. These only represent about one percent of all the stars, and the best example is Alpha Centauri A. In fact Alpha Centauri is a triple star system - The G-type Alpha Centauri A and the K-type Alpha Centauri B orbit around each other, whilst the red dwarf Proxima Centauri, the closest to us, orbits around both of them, but very much further away. It is a good example because not only is Alpha Centauri our nearest neighbour, (about four-and-a-half light years away) but it is comprised of three different kinds of stars.

K-TYPE STARS

The next least numerous stars are orange K-type dwarfs. These are hotter and brighter than red dwarfs. One of them, Alpha Centauri B, again one of our closest neighbours, is a good example. It is slightly smaller, dimmer and cooler than our own star, the Sun. These kinds of stars account for about 15% of all stars in the sky.

RED DWARFS

The most common and numerous stars are the ones called red dwarfs. These are small and glow dimly with a reddish light. They are so dim, small, and comparatively cool, that even the closest star in the sky, Proxima Centauri, cannot be seen with the unaided eye. You'd need at least a telescope to spot it. These red dwarfs account for about 70% of all stars in the Galaxy.

WHITE DWARFS

Another ten percent of all stars in our Galaxy are white dwarfs. These are the cores of old burnt out stars that have puffed out most of their material into space, and all that is left is the still shining heart of the dead star, sometimes surrounded by a "smoke ring" of its lost atmosphere. Early astronomers called these "smoke rings" planetary nebulae, although they have nothing whatever to do with planets. The Ring Nebula in the constellation of Lyra is among the best known examples of a planetary nebula. It is also known that Sirius (or the Dog Star), the brightest star in the sky, has a white dwarf companion, Sirius B. The stuff white dwarfs are made up of is very dense, and a teaspoonful would weigh about as much as a lorry-load full of sand. This is because the dead star's matter is very highly compacted (much like squeezing a slice of bread down to pellet size, only much more so).

BLACK DWARFS

Although not yet seen, these are thought to be the remains of white dwarf stars that have grown so cold they no longer have the ability to shine.

BROWN DWARFS

These have also never been directly observed and would be in size somewhere between a large planet like Jupiter, and the smallest star. They are reckoned not to have amassed enough matter to "ignite," 'thermonuclearly' speaking, and so radiate light and energy; therefore, they can't be seen optically (although some astronomers claim to have found evidence for their existence using highly advanced infrared techniques).

However, scientists have estimated that there should be more mass in the Universe than there is, and have developed this theory of brown dwarfs to account for this "missing mass."

GIANTS AND SUPERGIANTS

These are so large that some of them could swallow up our own Sun hundreds or even thousands of times. They are described by their hue as red, yellow, white or blue giants or supergiants. Some are actually stars going through their old age for, when a star like ours has exhausted its hydrogen and helium, it bloats and swells up to hundreds of times its original size and becomes a red giant. Larger red supergiants are the familiar Betelgeuse in Orion and Antares in Scorpio. Examples of blue and white and Supergiants that shine like beacons in the night sky are Rigel in Orion, and Deneb in Cygnus. Other well-known red giants are Arcturus in Boötes and Aldebaran in Taurus.

As large as they are, the matter they are made up from is very light, and a volume as big as a house of say, Betelgeuse, would only weigh about the same as a teaspoonful of water. Some of these supergiant stars are very, very large, and to compare our own Sun to say, a supergiant like Antares (in Scorpio) would be like comparing a frozen pea to a pumpkin! So it's no wonder they shine so brightly in the night sky, even though they're so very far away.

Having now accounted for most of the star types in our Galaxy, we are only left with a few remaining and more unusual ones.

PULSARS OR NEUTRON STARS

These peculiar stars are reckoned to have been created when a supergiant star ends its life, not more peacefully - such as happens when main-sequence stars like our Sun have used up most of their nuclear fuel and puffed their atmosphere into space to form a pretty planetary nebula in the sky - but in a highly spectacular explosion called a "supernova." The star explodes, whilst at the same instant its core collapses into a very, very dense object called a "neutron star" (a star made up mostly of neutrons) or a rapidly spinning neutron star, called a "pulsar." The core of the supergiant star is so compressed and compacted that in size it would measure only a few miles across - about the size of a city, but a lump of its matter, the size of a small sugar cube, would weigh as much as 20 million elephants!

So odd are these pulsars that quite a stir occurred in 1967 when radio astronomers at Cambridge picked up pulsating radio signals and thought these might be signals from other intelligent beings in outer space! It turned out to be a pulsar - the first one ever discovered. These pulsars spin very rapidly, at the same time flashing on and off, sending beams of radiation from their poles into space, like a celestial lighthouse. Although many pulsars have now been discovered, the pulsar in the Crab Nebula is, perhaps, the most famous, and was the result of a supernova explosion in 1054, just twelve years before the Norman conquest of England. It lies 5,000 light years away from us, and the remnant of the supernova - a misty patch - is still a fascinating object for viewing by professional and amateur astronomers alike. It is the last supernova event in our galaxy and has been mentioned in ancient Oriental writings.

The reader can be reassured that such supernova events in our own Galaxy are extremely rare, happening about once in a thousand years.

BLACK HOLES AND THE WARPING OF SPACE

Among the most bizarre objects in space are "black holes." You have probably heard much about these on TV documentaries. Indeed, they seem to dominate the subject of every astronomical chat-show. But what, you may well ask, are these peculiar objects? Well, to try and describe them briefly, they occur in the aftermath of supernova explosions, when the central core of an exploding star collapses even beyond that of a neutron star. It could wind up the size of a frozen pea, or even less!

The enormous gravity that these objects exert is so much that even light can't escape from them. For example the Earth has an escape velocity of about seven miles per second, which means that a rocket has to reach this speed before it can escape from Earth's gravity into outer space. Well, a black hole would have such an enormous gravitational pull that even light, at the speed of 186 thousand miles per second, is not fast enough to escape; hence these objects can never be seen. However, strong radio sources, in the form of x-rays, have been detected as emanating from very small but massive objects deep in space, and it is generally accepted by most scientists and astronomers that these come from black holes, which indeed exist. The concept of black holes is not new; in fact they were first suggested by John Michell way back in 1783, but forgotten about, then revived by Albert Einstein in the early part of the 20th century.

Gravity actually warps the space around it. This is a bit difficult to describe, but if you imagine placing a heavy object, like a bowling ball, on something like a trampoline, you will soon notice that the bowling ball has caused a "dent" in the trampoline. If you were to roll a marble slowly on the trampoline, it will roll down to the bowling ball, not because it is attracted to it but because the trampoline (our outer space) is warped by the weight of the ball. This is what is meant by the "warping of space." Of course, in our the illustration of the bowling ball and trampoline we are thinking in two dimensions, whereas this effect really happens in three dimensions, which is difficult at the best of times to simulate, or even perhaps to imagine!

This warping of space is most prominent around a black hole - it would be rather like imagining that our trampoline has such a heavy object sitting on it, that it has created a stocking-like dent in the trampoline material. The neck of the stocking would be very much like the "event horizon" which surrounds a black hole, and once you fell into it, you would never have any hope of ever being able to climb out again! Some theorists even suggest that this warping of space could continue in a "runaway" fashion, and that anyone or anything going in would exit in another universe! Some of these black holes are thought to be orbiting another star, "sucking" stellar material into itself and causing the other star to lose its mass.

QUASARS

These quasi-stellar (star-like) radio sources are very energetic and distant objects that are thought to be compact regions in the centres of galaxies with a supermassive black hole. Though comparatively small – only about the size of our Solar System – they radiate as much energy as about 1,000 times the amount of stars in the Milky Way put together, making them the most distant, luminous and energetic objects in the Universe.

These supermassive black holes are gobbling up the surrounding stars in a feeding frenzy, and ejecting energetic jets of radiation from their cores for thousands of light years into space. It is sometimes the sad fate of stars that instead of being allowed to end their lives naturally, they are instead devoured by a nearby black hole or a supermassive one.

Like life on Earth, life in the Cosmos too has its diversity of structures, shapes and forms, and goes through the processes of birth, life and eventual death, either normally or more dramatically, just as we do. The Universe certainly seems to behave like a super-intelligence, governed by the laws of physics, self-regulating and in complete balance. Perhaps the concept of the ancient Chinese, who believed that the Universe was an organism, was not so far wrong after all!

[19]

Galaxies

BEFORE EDWIN HUBBLE'S ARRIVAL ON THE SCENE AT THE MOUNT WILSON TELESCOPE IN CALIFORNIA, IN 1919, IT WAS THOUGHT THAT OUR UNIVERSE CONSISTED OF ONLY ONE GALAXY AND THAT WAS OUR OWN MILKY WAY.

Hubble changed all that, because by careful observation through the years, he discovered that nebulous patches of light amongst the stars were not nebulae at all but in fact other galaxies beyond our own. This was a breakthrough event because it increased the size of our Universe a billion-fold.

As he observed, there are in fact billions upon billions of other galaxies outside our own that each contains thousands of millions of stars. Imagine you were a bug living on the leaf of a leafy tree which represented your position in the Universe, the Solar System. All the other leaves on your tree would represent the stars in your galaxy. But then, look beyond your tree and you would see thousands, upon thousands of other trees, because you are in the middle of the Amazonian rain forest, and each of those trees would represent another galaxy.

There would be trees, trees and then more trees, stretching from horizon to horizon. That gives you some concept of how infinitely large our Universe is, and how many galaxies there are.

It is no wonder then, that Edwin Hubble was awarded the Nobel Prize for Physics, albeit after his death in 1953. His namesake telescope, The Hubble Space Telescope, or HST, has observed and taken spectacular photographs of other galaxies such as the Andromeda Galaxy, which is our closest neighbouring galaxy. It lies over two million light years from us and is the furthest object that can be seen by the naked human eye. With its far-seeing eye, Hubble was recently pointed at an apparently empty part of the night sky, and what did it find? Again, thousands upon thousands of galaxies, some so far away as to almost the beginning of time.

Edwin Hubble also observed that galaxies moved away from each other at incredible speeds, by calculating their redshifts. That is to say that a faster a galaxy moves away from you, the more its spectrum is shifted towards the red. This is known as the Doppler effect.

TYPES OF GALAXIES

Galaxies come in various shapes and sizes. Spiral galaxies, barred spirals, like our own and Andromeda, elliptical and others. Most are thought to have a supermassive black hole in their centres, which cause them to rotate and sometimes emit jets of energised particles that shoot out hundreds of light years into space.

ELLIPTICALS

According to the Hubble classification system, these ellipticals are ranged from EO (almost spherical – like looking at the lens of a magnifying glass full-on) to E7 (highly elliptical – viewing the magnifying glass lens edge on). They don't seem to have much structure in appearance and little interstellar matter, like gas and dust. These galaxies, consequently, don't seem to have many open clusters, and there doesn't seem to be much activity in them, like new star formation. They seem to contain more older stars, generally and are more like the globular clusters that hang over our Milky Way. The largest galaxies are giant ellipticals. It is thought that some of these galaxies are the result of a collision or merger of two galaxies into one so can grow to enormous sizes.

UNBARRED SPIRALS

Spiral galaxies are constituted of a rotating disk of stars. According to the Hubble classification scheme, these spirals are listed as type S followed by a letter a, b or c to indicate how tightly their spiral arms are wound up, and the size of their central bulge. An Sa would have tightly wound up and poorly defined spiral arms and have a relatively large central core, whereas, at the opposite extreme an Sc type galaxy, would have wide open and well defined arms, harbouring many bright young stars, and a small central core. The Whirlpool galaxy is a good example of an unbarred spiral.

BARRED SPIRALS

Barred spirals, classed SB followed by a lower-case a, b or c, are more numerous than the unbarred variety. These have a bar-shaped central band of stars running out of the central core that merge into the spiral arms. These bars are thought to be the result of a density wave travelling outwards from the core, or else some sort of "tidal interaction" with another galaxy. Many of these barred spirals are thought to be active. NGC 1300 in the constellation of Eridanus (The River) is a good example of a classic barred spiral, and it lies 50 million light years away from us. Our own Milky Way galaxy is a barred-spiral, classified as type SBc.

LENTICULAR GALAXIES

These are categorised as Hubble type S0, for the unbarred type, and SB0 for the barred. They are an intermediate form of galaxy that has the properties of both elliptical and spiral galaxies. They have ill-defined spiral arms with an elliptical halo of stars. NGC 5866 The Spindle galaxy in the constellation of Draco is an example of a lenticular galaxy.

PECULIAR GALAXIES

These galaxies have peculiar shapes and don't fit into any particular category of galaxies. Their strange shapes seem to be the result of tidal forces that happen when they interact with another galaxy. An example of a ring galaxy is Hoag's Object, which has a ring-like structure of stars surrounding a bare core. A ring galaxy is thought to occur when a smaller galaxy passes through the core of a spiral galaxy.

DWARF GALAXIES

These are far more numerous than the other types of galaxies we have discussed, and seem to occupy most of the visible Universe. These dwarf galaxies are much smaller by comparison to galaxies like our Milky Way, being only about

100th the size. Many of these dwarf galaxies orbit a larger galaxy, and it is reckoned that the Milky Way has at least a dozen such satellite galaxies.

Besides the types of galaxies mentioned above, there are a number of galaxies that do not fit in with the Hubble classification. Irregular galaxies do not seem to have any particular structure and our own Magellanic Clouds that hang over our Milky Way are examples of irregular galaxies.

A study of 27 of our Milky Way neighbours revealed that the central mass of all dwarf galaxies is about 10 million solar masses, regardless of whether the galaxy has thousands or millions of stars, indicating that there is something else, besides gravity that is holding the galaxy together.

DARK MATTER

This is a much discussed subject in all scientific circles. Simply put, it is matter that is thought to exist to account for the apparent "missing mass" that the Universe should have. Since it can't be seen like bright objects such as stars and the nebulae they illuminate, dark matter is invisible and is hypothesised to exist. Without dark matter holding galaxies together, it is theorised that their spiral arms would be thrown off into outer space.

DARK ENERGY

There is another force or energy that scientists are speculating over, and that is something called Dark Energy (not to be confused with Dark Force as in the Star Wars films).

It is hypothesised that Dark Energy is a form of energy that permeates all of space and tends to accelerate the expansion of the Universe.

An alternative explanation is that the current Big Bang theory is actually wrong. Dark Matter and Energy are theoretical concepts to make the model work. Like many aspects of this book, our understanding will change over time and this will certainly not be the last edition!

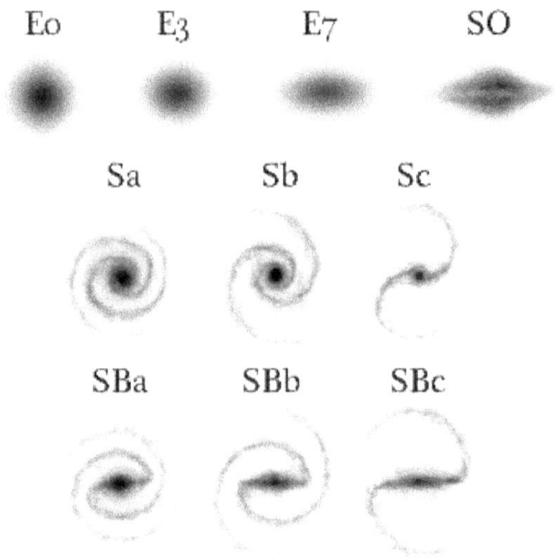

[20]

When Time Began

IT IS ONLY SINCE THE 20TH CENTURY, WHEN WE BEGAN TO STUDY THE UNIVERSE WITH MORE AND MORE POWERFUL TELESCOPES, RADIO TELESCOPES AND OTHER ASTRONOMICAL APPARATUS, THAT WE BEGAN TO COMPREHEND IT AND HOW IT MAY HAVE STARTED, HOW BIG IT WAS AND HOW IT WOULD END.

We know now that it is very, very much bigger than we ever imagined and that it appears to be getting bigger every minute with expansion. We have probed almost to the Universe's beginning by observing bright objects like quasars that seem to lie at the fringes of an infinite Universe, and therefore to almost the beginning of time. Not only has our knowledge of the Universe increased with the vast amount of data we have gathered through our scientific apparatus, it is with the mind that that we are now beginning to comprehend the Universe – such minds as Einstein, with his theories on relativity, and Hawking's concepts on quantum theory. Data itself is not enough; it is the interpretation of the data that is so important, and brings to mind the little children's rhyme "Twinkle, twinkle little star. How I wonder what you are."

EARLY CONCEPTS OF THE UNIVERSE

In ancient times, and perhaps even to this day, not much thought was given to the Universe or how it came into being, and Judo-Christian religious belief always held sway that in the beginning it was void and without form and created divinely in six days. This was calculated to have started only 6,000 years ago. Other religions considered that the Universe was closely linked to the actions of gods and goddesses and that the Heavens were always the preserve of these gods, Earth the preserve of man, and Hell the destination of the wicked. At times these gods came down to Earth and interacted with humans. Prayers and sacrifices were offered to these gods to appease them, and seek grace and favour. The ancient Greeks did have the concept that the Universe had been a motionless homogeneous "mixture" that later turned into a whirling mass into which cold matter fell towards the centre, forming the Earth. The Sun, Moon and planets were thought to have been torn out of the Earth. The ancient Chinese considered that everything on the Earth and in the sky was part of a gigantic organism, and in ancient India, they believed the Universe underwent a cycle of destruction and rebirth, each cycle taking millions of years.

THE STEADY-STATE UNIVERSE

Back in the 1940s, the Steady State hypothesis was the previous model for the Universe, but by the end of the 1960s this theory was largely abandoned, and though there are still some scientists who argue their case for a "Steady State Universe," evidence that the Universe is expanding – they use the term "inflation" rather than expansion – is making their argument weaker.

Their hypothesis was that new matter was being created and lost quietly and continuously – much like a bowl in the kitchen sink being constantly filled with water from the tap, and allowed to overflow at the same rate. This concept would represent a steady-state Universe of constant regeneration, the theory being that while material was being swallowed up by black holes, the matter emerged again at the other end in so-called "white holes." Quasars were wrongly thought to be the exit point for this outpouring of matter. The Big Bang theory is now the accepted model of the Universe.

THE BIG BANG THEORY

Who can say they have not heard of The Big Bang? It is difficult to avoid, as it is thrown at you at almost every scientific TV program that is being screened. But to hear is one thing; to understand, quite another! Simply put, it is the beginning of the entire visible Universe, and time and matter itself. According to physics, at time zero there was an infinitely hot, infinitely dense and infinitely small point of energy called a "singularity." This point would fit many times into the period at the end of this sentence. This singularity sort of "exploded" and created all matter, time and space in an instant. This event took place about 13 billion years ago, and according to two scientists, Arno Penzias and Robert Wilson and the horn antenna they designed and built, which resembled a giant ear-trumpet, revealed the background microwave radiation left over from the Big Bang that can still be detected.

If you ever want to see this residue radiation for yourself, you don't need a complicated horn antenna; just switch your TV to an untuned channel, and the dancing white dots of light across the screen will display this for you more than adequately! With the Big Bang, time, space and all the matter within it, "inflated" much like a balloon, so the galaxies are moving apart from each other at a rate of knots (figuratively speaking). Actually, they are not physically moving, but moving with the inflation. If you had a balloon and drew a few dots on it while it was flat, then inflated it, the dots would move further apart – not because they are physically moving away from each other, but that space itself – in this instance the material of the balloon – is expanding

MULTI-UNIVERSES

Perhaps there is not simply one universe but we are just one of many billions of universes being continuously created and destroyed, much like the water in a saucepan that is slowly being brought to the boil, where tiny bubbles form at the bottom from apparently nowhere only to disappear instantly, when they reach the top.

Studying a map of background cosmic radiation of the Universe and comparing it with data gathered by the Planck spacecraft, cosmologists have determined that the anomalies revealed can only be explained by the gravitational pull of other universes, and they conclude this is the first hard evidence of the existence of other universes. The eminent Stephen Hawking is a firm supporter of the theory of multiple universes.

HOW WILL IT END?

Having now established that the Universe was created with the Big Bang, you may well ask, how will it all end?

At one time, cosmologists thought that after aeons, the Universe would stop expanding, and all the matter within it would start to come together again in something they termed "The Big Crunch," Eventually this matter would contract to a point where it became another singularity, and the process of another Big Bang would follow. However, cosmologists now seem to agree that there is not enough matter, and consequently gravity, to overcome the rate of expansion to allow this coalescence to happen, and as such, the Universe will continue to expand indefinitely, eventually to become so spread out as to disappear altogether into infinity. But of course, no one really knows, and we will never be around to witness the event.

Although the Universe, as we see it, may appear to us as serene and tranquil, this is anything but true. It can be a violent and very dangerous place, with stars exploding or being gobbled up by neighbouring black holes, lethal gamma rays, x-rays and solar flares. Were it not for the protection of our magnetic field even our Sun's solar wind could sweep away our precious atmosphere in a trice. Fortunately too we are far enough away from deadly stars, like Eta Carina (9,000 light years away) which has two enormous lobes of distended gas at each end, and which may go supernova at any moment (if it hasn't done so already). If a nearby star exploded, we would not have much chance of surviving the incident. Perhaps, after all, it's not such a bad idea that the Universe is so BIG. Distance protects us not only from hostile stars but perhaps from hostile aliens too!

[21]

Science Fiction : Science Fact

SCIENCE FICTION WRITING REALLY GOT STARTED TOWARDS THE LATTER HALF OF THE 19TH CENTURY WITH WRITERS LIKE JULES VERNE AND HG WELLS.

Early sci-fi writers confined their writings perhaps closer to home, or to home itself. The Moon and the nearer planets were the subject of most of their material, and by the mid-1930s "Martians" became a popular name for aliens from Mars. Some of "you ancient ones" might still recall the panic caused by Orson Welles when he simulated a news broadcast based on "The War of the Worlds" in 1938 to the effect that our planet was being invaded by Martians!

As mentioned earlier, Percival Lowell with his observations of canals on Mars, lent popularity to the idea that this planet might be inhabited by intelligent beings. Reports of flying saucers after World War 2 also lent credibility to the possibility of intelligent life on our neighbouring planets. This was all good stuff for science fiction writers. But this was short-lived when space exploration took off and we actually landed men on the Moon in 1969, and sent spacecraft hurtling off to the inner and outer planets in our Solar System.

So sci-fi writers began to look further afield and stretch their imagination to the stars – after all, why bother with the "boring old Solar System" where no aliens lived. "Science faction" also began to get boring to the general public. Why bother to see astronauts on television driving around the Moon on their moon-buggy when it was much more interesting to see what Kirk and Spock were up to "boldly going where no man had gone before" in the latest episode of Star Trek. Ingenious ideas were dreamt up by sci-fi writers of how to flit from star to star in the wink of an eye, and interstellar travel problems were solved by the "invention" of hyper-drives, warp drives and the like to enable these inter-stellar travellers to far exceed the speed of light.

Problems like zero gravity were not touched upon nor the reality of actual space distances. Aliens of all kinds were dreamt up, and these ranged from Earth-like beings, to insect, animal and reptilian conceptions. There are no holds barred to the imagination of sci-fi writers and film and TV producers!

However, as far-fetched as science fiction appears to be, some very novel and ingenious ideas have emanated from various sci-fi books, films and TV series. I have picked out a few of the more popular ones as examples as what we imagine as fiction has the habit of forming our new reality.

2001 A SPACE ODYSSEY

Perhaps one of the greatest science fiction books ever produced was Arthur C. Clarke's "2001 A Space Odyssey" and the film made by Stanley Kubrick. This at least tries to resolve some of the problems of space travel, even if it was within our own Solar System, although, as it turned out, 2001 was an optimistic year for such advanced technology. Some who may have seen the film when it was first released at the end of the sixties left the auditorium with an air of bewilderment, unable to understand the film. But my own personal feeling is that is just what it meant to do – not attempt to answer the unanswerable.

CONTACT

Since then, Carl Sagan's excellent "Contact" was published. Since he himself was a distinguished astronomer his book was significantly one notch above the usual science fiction books and films that we are more familiar with. It deals with the concept of "worm-holes" in space – i.e. the warping of the fabric of space itself to create a shortcut to the stars, and this is an idea that has been attractive to scientists who would like to cut the journey to the stars to a fraction. It is also about the SETI project (of which Carl Sagan was a firm supporter) that started receiving coded signals from outer space containing the plans for building a machine that could create a wormhole in space. The film also tackles the 'science versus religion' debate, but concludes on a note that suggests that science and religion can co-exist rather than be at loggerheads. Carl Sagan died in 1996; before his story was made into a film of the same name in 1999, starring Jodie Foster.

STAR TREK

A very popular sci-fi TV series by Gene Roddenberry that originated in 1966 and in later years was followed by many TV sequels, films and spin-offs. It has a cult following of "Trekkies" who dress in Star Trek gear and faithfully attend every possible function, sometimes travelling halfway around the world to do so. The series has also generated some ingenious ideas that have inspired the invention of devices like the mobile phone, morphed from the concept of the communicator. Similarly, video-communication is now a reality with face-to-face interaction on Skype. These could only be dreamed of 20 or so years ago. MRI and CAT scans are being used to produce computer images of the human body, much like the tricorder in Star Trek was used to analyse medical conditions.

Even simple ideas like automatic doors which magically opened and closed for the Star Trek crew, are now in virtually every supermarket. Voice commands used to instruct the Enterprise computer are now being used in mobile phones, and portable storage devices, which Spock casually slipped into the ship's computer, are now commonplace with USB memory sticks and memory cards on digital cameras. Not only did Star Trek inspire new technological ideas, but it also included social issues, like the first multicultural cast in the ship's crew, and televised the first inter-racial kiss!

DOCTOR WHO

This flamboyant and slightly eccentric Time Lord travels through time on his "TARDIS," a time-travelling space ship, which appears magically as a blue British police box from the outside, but changes its interior to a space-time vessel once inside. The Doctor meets various challenges in his time travels, including the dreaded Daleks with their cries of "Exterminate! Exterminate!" The series is the longest running BBC TV series that has been aired from 1963. As such, various actors have played the part of the Doctor, and this fits into the scheme of things, as a process of "regeneration," - a life progression applicable to Time Lords – takes place, and they can take on a new body, and sometimes a new personality, should the incarnation be injured or damaged in some way.

CLOSE ENCOUNTERS OF THE THIRD KIND

This film released in 1977 and directed by Steven Spielberg, is based on the UFO question. The story is of aliens who seem to wish to establish close contact with us humans, and implant images of the place of rendezvous (Devils Tower, Wyoming) in certain chosen individuals, and also a signal of five musical notes.

These individuals (one of whom is actor Richard Dreyfuss), become obsessed with these images, and are drawn to the meeting-place where eventually the enormous mother ship lands. The aliens establish contact with a team of scientists who are ready and waiting to receive them. The aliens return certain abductees who have been taken by them over the decades and more recently, including a five-year-old boy and the pilots of missing Flight 19 that mysteriously disappeared over the Bermuda Triangle in 1945. They invite on board their mother ship, volunteers they have chosen to take back with them to whatever alien planet they came from.

BACK TO THE FUTURE

Directed by Robert Zemeckis and released in 1985, this film is about time travel. Our hero, a young teenager, played by Michael J. Fox, is accidentally sent back twenty years in time to before he was born, in a time machine, invented by his professor friend, in the shape of a DeLorean car. Here he meets his parents-to-be, still in high-school, but unluckily his mother is attracted to him in preference to his father. In order to prevent such interference in the space-time continuum, he has to make an effort to get his parents to fall in love and marry, so that he can continue to exist in the future! He manages to accomplish this with the help of his professor friend, who he meets up with again in the past, and returns again to his own time. This is an interesting film because it raises the old paradox question of whether travelling back in time is at all possible.

STAR WARS

Here we are in the realms of science fantasy rather than science fiction. This film even goes beyond our Milky Way galaxy and starts with an almost fairy tale phrase "In a galaxy far, far away,." The story line deals with the forces of good, as depicted by the Jedi knights, against the dark forces of evil, represented by the Sith and headed by Darth Vader, kitted out in a suit of dark armour, inspired, no doubt by that of a samurai warrior. Here the forces of good and evil clash in a display of fencing with light sabres. All sorts of strange planets and stranger creatures feature in these series of films, characters like Jabba the Hutt, (a slug-like creature), Chewbacca (a Wookiee that stands 7 feet tall and resembles a hairy bear-dog) and two robots, R2-D2 and C-3PO.

The heroes are the Jedi knights, led by Alec Guinness playing the part of Obi-Wan-Kenobi, and a strange little creature called Yoda. All kinds of robotic war machines and vehicles were dreamt up. The films, starting with Episode IV, V and VI, and their prequels I, II and III were directed by George Lucas, and were the blockbusters of their time.

THE TERMINATOR

Directed by James Cameron and released in 1984, this film deals with cyborgs – half-man and half-machine that have been developed in the distant future. The cyborg, (in the shape of Arnold Schwarzenegger, famous for his "I'll be back" phrase in the film), is sent back to the past (our current day), as an assassin to terminate the heroine, Sarah Connor, the mother of a future rebel leader. The original film was followed by three sequels and a TV series called The Sarah Connor Chronicles. The concept of half-man and half-machine has been used as the theme of some other sci-fi films too.

[22]

Time and Space Travel

TRAVELLING IN TIME HAS BEEN POPULARISED BY MANY AUTHORS THROUGH THE AGES.

Probably the most significant and seminal work is from HG Wells. In The Time Machine, written back in 1895, the hero builds a time machine and travels forwards far into the future as well as back in time. Travelling backwards in time is also the subject of more modern and futuristic books and films like Back to the Future.

The concept of moving forward in time seems easier to comprehend and is thought more feasible, scientifically, than travelling backwards in time which is always confronted by the uncomfortable paradox, "If I were to travel back in time and kill my own grandfather, would I, in fact, exist today?"

We are in fact travelling through time all the time. We are locked in a moving wave of time, and not only us – the whole Universe, ever since the Big Bang, is locked in this time-trap, this cosmic timepiece which progresses as steadily as the ticking of the clock on your own mantelpiece.

But what if time was not like that, and instead of being a single unfolding history, was more like a many-branched tree, each branch representing a different history of time. If then it were possible for someone on a growing branch of the tree (his time), to do a "quantum leap" back to his past when he was a child, could he then exist twice; as himself and as another – his child self? However, our time-traveller would have to be extremely careful in this scenario, for if he created an event which could cause a disturbance in the space-time continuum – say he was responsible for an event which caused an accident resulting in the death of his child-self – another "branch" of time would begin to "grow," and he would be in an alternate time-line, where his child-self would have died, but he, himself could exist in this new time-world he had caused.

Then if our intrepid time-traveller wished to return to his own time-line, he may have to consider returning to the point where he interfered with the space-time continuum,* thus undoing the damage he did, and then from there return to the point of time from where he originally started, with probably no time having passed for him at all!

This may avoid the awkward paradox effect which has been the sticking point and plagued all scientists to this day causing them to consider that travelling back in time is not possible at all!

Such a theory, "The Many-Worlds Interpretation", does exist, where such branching or splitting of time happens, creating as many different timelines as alternative events that might ever have happened throughout our history, events like say – Hitler's father had decided not to sleep with Hitler's mother on the night of his conception, would the Second World War never have happened? Every cause has an effect, and as we weave our way through life, every decision we make – however simple – can change the course of history in our time. So in this "Many-Worlds" concept, every alternate decision we make would open another time-line to a different history. If you consider the lifetime of every individual who has ever existed on this planet, and the different choices they have made in the course of their life, there could be billions upon billions of such parallel Universes, each representing the possible outcome of every event in its own particular history.

This theory of matter existing in more than one place in more than one time is one that is being discussed and debated in scientific circles, based on the principal that some atomic nuclei can exist in more than one place at one time simultaneously, so the idea is not completely in the realms of science fiction. Stephen Hawking himself is a supporter of "The Many-Worlds Interpretation".

The vast distances of interstellar space make it seemingly impossible that one can travel to other stars with the ease depicted in various sci-fi films and books. Unfortunately, the time it takes for even light to travel from star to star at 186 thousand miles per second is just too slow! It is almost a snail-like crawl, or the speed of a hair growing, in relation to the sheer size and vastness of space.

The only concept of covering these unimaginable distances is to conceive of taking a shortcut, and that is where the theory of wormholes in space emerges. If you were to imagine you were holding a balloon in your hands and then drew two dots at opposite sides of the balloon's surface representing the points of the distance you wished to travel, then instead of drawing a line over the surface of the balloon to join the two dots you, instead, pushed your two fingers and squeezed through the balloon until your fingertips met, you would have created a wormhole – a short-cut through space! In other words, you have warped the surface of the balloon, representing space-time, creating a time-warp. Extreme forces of gravity can create this warping of space – forces that are only found in the vicinity of a black hole.

According to Einstein's theory of relativity, It is also conceived that the faster you travel, the more time slows down, as time and space are intrinsically connected – hence the term "space-time." The speed of light through the vacuum of space travels at a constant 186 thousand miles per second, and cannot go faster; therefore, if anything is perceived to slow down, it has to be time!

If a spaceship were able to reach about 90 percent the speed of light, the time on the spacecraft would slow down, so that an astronaut travelling for a year at that speed would only have aged a year, whereas according to Earth-time, a hundred years may have gone by, and he would return to a world where all his friends and family would have died. This slowing down of time in relation to speed can be measured by synchronised atomic clocks in fast-flying aircraft, but it is so minuscule as to be hardly noticeable at all.

Approaching the event horizon of a black hole would also create this effect of the slowing down of time, as strong gravitational forces have the effect of slowing down time. In fact, if it you were able to witness a spacecraft approaching the event horizon of a black hole, it would appear to hang there forever, as time would have slowed down so much it would not appear to move to an outside observer.

[23]

Is Anybody There?

THE QUESTION OF WHETHER WE ARE ALONE IN THE UNIVERSE HAS HUGE IMPLICATIONS NO MATTER WHAT THE ANSWER MAY BE.

From "The First Men in the Moon," by author HG Wells in 1901, we have been seeking other life forms. First on nearby objects like the Moon, and then further afield like Mars and Venus, constantly moving the goalposts to the outer planets, desperately seeking alien companions! As all the planets in our Solar System have proved uninhabitable we have now had to turn our eyes towards the stars, and only those that have been found to have some sorts of planets! Given that under a friendly sun we have 9 planets in our immediate neighbourhood and several terrestrial moons and not found one sign of life – not even an amoeba – were we to find some evidence of alien life somewhere, given the vast distances of outer space, would it really matter? Our Pioneer spacecraft have been dispatched with messages telling anything lucky enough to find them that "We are here," and although SETI (Search for Extra-Terrestrial Intelligence) has been eavesdropping on our skies since the early 1960's "not a sausage" has been found!

However, in 1961 Dr. Frank Drake of the University of California, astronomer and astrophysicist devised an equation on the possibility of other intelligent life in the Universe.

$$N = R^{*} . f_{p} . n_{e} . f_{\ell} . f_{i} . f_{c} . L$$

Where:

N = the number of civilisations in our galaxy with which communication might be possible

R* = the average number of star formation per year in our galaxy

fp = the fraction of those stars that have planets

n = the average number of planets that can potentially support life per star that has planets

fl = the fraction of planets that could support life that actually develop life at some point

fi = the fraction of planets with life that actually go on to develop intelligent life (civilisations)

fc = the fraction of civilisations that develop a technology that releases detectable signs of their existence into space

L = the length of time for which such civilisations release detectable signals into space.

He concluded from this estimate that there are approximately 10 thousand planets in the Milky Way that might contain intelligent life with the possibility of communicating with us.

Here one has to speculate. Given the billions and billions of stars that exist in our own Galaxy, leave alone the billions and billions contained in billions of galaxies that can be seen with the amazing eye of the Hubble Space Telescope, it is more than possible that intelligent life does exist somewhere in this vast and seemingly unending Universe. But there are various possibilities to be considered.

A CLOUD COVERED WORLD

Just suppose our nearest neighbours lived in a world obscured with clouds – much the way Venus is – deprived of the opportunity of observing clear and starry night skies. We have been lucky; for aeons our ancestors have observed the stars and planets under clear and starry skies and tried to work out what they were. Not so our unlucky alien neighbours! Clouds are all they would see, and if they ever achieved the curiosity and technology to climb above the clouds, they would see something that they had never known existed – a clear and starry Universe. So, we have a great head-start on our unlucky stellar neighbours!

A WATER WORLD

We know that intelligent creatures, like dolphins, exist in our vast oceans, and an underwater civilisation is a very romantic concept. But like our cloud-covered neighbours above, they too would be largely unaware of the star-filled night skies, unless they came to the surface occasionally, and that would not give them sufficient time for observing the night skies like ourselves. Another head start for us humans! Another disadvantage for our watery neighbours: in an underwater world with no land, fire – that that ancient source of light and energy – would not have been readily available to forge any sort of primitive tools. It would have been a struggle for them to achieve any sort of technological advancement, unless of course their world was anything like ours and had volcanic vents that could be a source of power.

AN UN-TECHNOLOGICALLY ADVANCED WORLD

No one can say that we were uncivilised even in the far and distant past; in fact we were a highly civilised, cultivated and advanced civilisation as recently as the 17 and 18 Centuries. We had travelled the globe, created great works of art, music and literature, but had not advanced technologically. All the power we had was generated by wind, water or muscle. Therefore, another neighbouring civilisation, at this stage of development, would not even be able to receive our radio or TV signals which we have been transmitting through space, and which would at least have reached any likely inhabited planet within 65 light years away. These are some points to ponder.

THE UFO QUESTION

Evidence of UFOs (Unidentified Flying Objects) cannot be denied. There are just too many eyewitness accounts from very reliable sources for them to be dismissed out-of-hand. But what are they? There is no definitive answer. All the hype began in 1947 with the "Roswell Incident," when a flying saucer was believed to have crashed on a ranch in Roswell, New Mexico, and has gone on ever since.

It has been the subject of documentaries, films and debates but with no positive or conclusive evidence of what these strange sightings really are.

Sometimes they are described as "flying discs," or "flying saucers," or other mysterious lights in the sky, sometimes flying in formation or hovering above cities. They have been observed by airline and air force pilots, and by astronauts themselves. The popular theory is that they are alien visitors from some advanced civilisation. If this is so, such a civilisation would have had to solve all the problems of space travel through enormous distances and the time that would take, even at the speed of light. Perhaps with their advanced technology they have managed to create a wormhole in space, from their part of the Universe to ours, and are therefore able to travel through this shortcut in space in a fraction of time.

There are also theories that these alien visitations are not new, but have been happening throughout our history. Whatever it is, these "visitors" have been very shy, and have left no real evidence of their visitations. There is another wild theory that these visitors are not aliens at all, but us thousands of years into the future, when we have found a means of travelling back in time in these craft. The change in our physical appearance, (if these sightings of alien-looking creatures are true), could be the gradual evolution of our species through the millennia. The most feasible explanation though, is that these strange sightings are of highly secret military developments, and conspiracy theories of government cover-ups abound.

The debate will continue, and someday, hopefully, we might come to a definitive conclusion.

[24]

The Shape of Things to Come

BEFORE WE TAKE A PEEK INTO THE FUTURE TO SEE WHAT POSSIBLE ADVANCES WE WILL HAVE MADE, WE NEED TO TAKE A LOOK OVER OUR SHOULDER JUST TO SEE WHAT TECHNOLOGICAL ADVANCES WE ALREADY HAVE MADE, PARTICULARLY IN THE 20TH CENTURY.

I'm certain that in future generations, the twentieth century will be referred to as the "Great Technological Leap Forward," for it is in this century that most of the advances of a modern civilisation have been made, and all practically within three generations.

We have leapt from Kitty Hawk to Concorde in under 70 years, landed astronauts on the Moon, and made such advances in medical science as to be able to replace someone's heart with another. One could go on and on, but I'm sure the reader will appreciate all the technological advances we have made in the last 100 or so years. After all, for centuries we were travelling the oceans in ships powered only by the wind, riding around in horse-drawn carriages and writing with quill pens. So much for technical advancement between the centuries! It was only in the late 19th Century with the start of the Industrial Revolution that we began to make any sort of technical progress. In my humble opinion, the advancement in the 20th Century was due mainly to a new power source – electricity.

I know most will like to remember the 20th Century as the Atomic Age, but after all, the awesome power of the atom has only been unleashed as a weapon of war and to power nuclear reactors, and in that latter role it is just a replacement for a coal-fired furnace to provide the heat to provide the steam to provide the power to drive the generators that produce what? Electricity! So "hats off" to the scientists who brought us into the Electronic Age, the real power behind the advancements of the 20th Century that has given us all our electronic wizardry including radio, television, computers, cell phones and the like. And where would we be without them today? It has literally lightened up our world from the dark age of candles!

Looking into my crystal ball, I am sure there will be many, many technical and scientific advancements in the near and far future, unless of course, we blow ourselves to bits in a nuclear holocaust or be sent back to the Stone Age after a meteor impact! I have picked out a few items from the hat, and they are:-

GENETIC RESEARCH IN MEDICINE

It is most probable that by the end of the century a cure for cancer, that great scourge of human health will be found. Probably this will be attacked at a genetic or cellular level rather than the crude means of the surgeon's knife. Also heart disease, another great scourge, will be eliminated, perhaps by growing a complete and working heart using stem cells. This could also be true for liver, kidney and other diseases of our vital organs. It is not impossible that new organs can be grown. After all consider the development of the human embryo where, in a matter of a few months, a complete human being is grown within the human body from the union of just two cells – the sperm and ovum. And it all happens by itself without us even having to think about it!

Nature has found the answers to such complex problems, the solutions of which we can only dream about. The metamorphosis of a caterpillar into a butterfly is another of Nature's many marvels, where the cells are completely redistributed within the chrysalis to form a completely different sort of animal. Lizards and some amphibians like salamanders and newts are particularly adept at growing new tails, limbs, jaws and eyes to replace ones that were lost. If you ever kept a tadpole in a goldfish bowl as a child, you will have witnessed on a daily basis its development into a frog. Even human and animal tissue has the ability to repair itself. Think of when you last cut your finger. So it's not impossible that in the future we will find a way of regenerating damaged limbs and organs using a genetic approach.

BEATING OLD AGE

Another probability is extending the human life span by slowing down the ageing process. Of course this would have all sorts of moral implications. Who would be chosen to live longer – the very rich, the very famous, the very brilliant? Like the alchemist's dream of turning base metals into gold, the "fountain of youth" has always been a dream that writers like Rider Haggard, author of "She," have exploited, though he used a flame of youth rather than a fountain of youth.

GENETIC ENGINEERING

Cloning another human being from the cells of another human being is a real possibility and has been done on sheep like "Dolly." But here too it will give rise to all sorts of moral implications. Would cloning be used by perhaps grieving parents of a child they had lost to replace it with an identical cloned model? Cloning has been used as the theme of some films like "The Boys from Brazil" where nail-clippings from Adolph Hitler were grown into several replicas of him. Would we really want a society of clones? Einstein, perhaps; but Hitler?

EXPLORING OUTER SPACE

It is also probable that we would have further explored the Solar System, perhaps landing robotic probes on Europa (one of Jupiter's moons), and drilling through the frozen surface to see if there is a liquid sea underneath, then sending down a robotic probe to investigate whether there was any form of primitive life living there. We may even have established a base on the Moon and landed astronauts on Mars. All this could happen within this century.

ANTI-GRAVITY MACHINES

Another possible invention in the future would be an anti-gravity machine. The concept is not new. It was used in HG Wells' "The First Men in the Moon" where the craft was made of a material he called "cavorite" which had anti-gravity properties. The capsule had windows with shutters all around, and when these were closed, the craft became gravity-free. Constructing such a machine would present awesome difficulties, as obviously we don't have any cavorite! However, it is thought that the Nazis in Germany were experimenting with such an anti-gravity machine during the Second World War to produce some sort of flying saucer which could defy gravity, much like the UFOs seem to do. Improbable as it may sound to produce such flying machines that overcome the effects of gravity (think of a flying car) is perhaps not impossible, and may happen within the next few centuries.

TELEPORTATION

"Beam me up Scottie" is the familiar phrase used by Captain Kirk in the Star Trek series to his Chief Engineer to beam him aboard the Starship "Enterprise." In the process he is "atomised" and transmitted to the Transporter aboard the spacecraft where he is re-assembled. It is indeed a very novel concept, but consider the technical difficulties involved in such a process. All the atoms in a person's body would have to be disassembled, beamed to the receptor, and then reassembled exactly as they were, including all the neurons in the brain! If it were at all possible to do this, the chances of something going wrong would be very real indeed, as depicted in the book "The Fly" by George Langelaan and later made into films in 1958 and 1986, where a fly somehow gets in to the transportation machine along with the human, and what comes out the other end is half-man half-fly! A very nasty consequence! So for the moment we will have to keep Teleportation in the realms of science fiction, where it belongs, and consider it a very improbable concept in The Shape of Things to Come.

[25]

Lost Technologies

IT HAS BEEN SPECULATED THAT OUR ANCIENT ANCESTORS MAY HAVE POSSESSED KNOWLEDGE OF CERTAIN TECHNOLOGIES THAT HAVE LITERALLY DISAPPEARED IN THE SANDS OF TIME.

One such case is Gobekli Tepe, a monumental stone structure that was buried 20 feet under the sand in South-Eastern Turkey, where it had lain undiscovered for 11,600 years, thus predating the building of the Pyramids at Giza by about 6,500 years. This ancient stone circle, built by an ancient Neolithic people, comprises of perfectly carved T-shaped pillars of rectangular stone, 18 feet tall, some carved with the figures of ferocious animals and other creatures.

This ancient megalith was built by a people who had not developed metal tools or even pottery. Also the mystery of the massive terraced walls at Saksaywarman, (Peru) where stone blocks, weighing some 50 – 100 tons, are so perfectly manoeuvred and fitted into place like a massive jigsaw, that the seams seem to be almost moulded together, as not even a sheet of paper can be inserted between them. How they managed to lift and move such massive stones with just muscle power is baffling archaeologists and scientists to this day.

Perhaps this ancient technology of moving huge rocks, apparently overcoming the force of gravity, was re-discovered by an eccentric recluse named Edward Leedskalnin, who was a Latvian immigrant to the USA. Back in the 1950s he devised some means to quarry, lift and manoeuvre huge blocks of coral, some 30 tons in weight, to construct the Coral Castle in Homestead, Florida, which he built in secret, single-handedly, without the help of any mechanical lifting equipment, except a small makeshift wooden tripod. He was a slight man, only 5 feet tall and weighing 7 stone, so how he moved and arranged these huge blocks, spread over several acres, without the help from anyone is a mystery to this day, and Leedskalnin took his secret with him to his grave. He claimed, in his journal, that he knew the secrets that helped the ancient Egyptians to build the Pyramids, to make the blocks into weightless objects.

ALIEN HELP?

There is also some speculation that alien visitors to our planet in the distant past helped in the construction of these massive stone building projects, like the Pyramids of Egypt, Stonehenge and the massive array of huge stones in Carnac, Brittany. They also reckon that these visitors helped design and build the Nascar Lines in Peru – those hundreds of ancient drawings on the ground of humming birds, spiders, monkeys and fish – that can only be aerially viewed from an aircraft. There also seems to be some sort of connection between these ancient megaliths to star constellations, for example the

Pyramids of Giza to the constellation of Orion, and the Newgrange Passage Tomb in Ireland to Cygnus. There may not be any truth in these conjectures, but one has to ask oneself, if ancient alien visitors had a hand in helping with these projects, what would be their purpose? It is a great mystery why our ancient ancestors went to all the trouble of using their time and muscle power to construct these gigantic structures. Was it for religious reasons or astronomical reasons or something else? One can only guess.

SCIENCE VERSUS RELIGION

Ever since the dawn of civilisation, religion has dominated human society. Every culture however primitive or advanced has worshipped some form of gods or deities. These beliefs are literally written in stone (like the Ten Commandments). Science, on the other hand is ever-changing and this has been a great advantage. As new discoveries are made, theories change, and science moves on. Religion on the other hand stayed put, and is now becoming out of step with the advances in scientific thought and technology.

The two philosophies seem poles apart and in constant conflict. This is a great pity, as both have played such an important role in the development of human society. Whereas religion is concerned with the spiritual, science is concerned only with the physical, and it is difficult to come to terms with both – the inner-self and the outer-self. Both are equally important. There is still a lot we do not know or cannot comprehend, and to throw one or the other out of the window and doggedly stick to one philosophy will not make us better human beings.

Some sort of compromise should be reached. Freedom of thought releases us from the shackles of any particular doctrine, and allows us to view the Universe with an open mind. The advancement of science has brought great benefit to mankind, and I am sure no one would like to return to "the bad old days", no matter what their religious persuasion.

[26]

Our Fragile Earth

POINTING ITS CAMERAS TOWARDS EARTH AS IT LEFT THE SOLAR SYSTEM IN 1990, VOYAGER 1 TOOK A PICTURE OF OUR PLANET WHICH APPEARS AS A MINUSCULE BLUE DOT AMONG THE STARS OF THE MILKY WAY.

This tiny speck is our homeland, a fragile bubble of life in a vast and, as far as known, uninhabited Universe. Everything that we hold near and dear is contained in this tiny bubble, and which, if it suddenly disappeared from the Universe, would not be noticed at all. It is often referred to as "the Goldilocks planet," because everything is "just right" for life, including ourselves, to exist. It is just the right distance from the Sun not to be too hot or too cold and the temperature to be just right for liquid water – the essential for all life – to exist.

Also it spins and circles the sun at the right speed and angle to give us a 24 hour day, and our seasons. There are sufficient land masses, rivers, lakes and oceans to accommodate the abundance of life found on every part of the planet, from micro-organisms to large mammals, reptiles, fish and plants. If we compare our Earth to a planet like Mars which is, sterile with not even a bacterium on it, with the abundance of life on this planet on every particle of dust or drop of water, we can understand how conducive to life our planet is. It seems we are uniquely "infected" with life.

The Earth is also a very efficient recycling plant. We lose a very negligible amount of our atmosphere to outer space, and every drop of water in our oceans, rivers and lakes evaporates into clouds and eventually finds its way back to earth as rain, hail or snow. All animal and plant material (and that includes us), decays after death, albeit with the help of bacteria, bugs and fungi, and not only turns to earth again, but "enriched" earth. Even rocks in the Earth's crust "sub-duct" back into the Earth's mantle and are spewed out again in volcanic eruptions as magma to make new crust. It is only "man-made" waste, like some plastics, that the Earth finds difficult to handle, and we have to intervene to turn this waste back into useful products. Otherwise everything made on Earth, stays on Earth, except some of the "space junk" we have orbiting around our planet or deposited on the surface of the Moon, Mars and elsewhere in space.

In some ways, our planet almost behaves like a super organism, looking after itself and keeping everything in balance. One only hopes that one day it won't get fed up with us and shake us off like a dog shakes off fleas! For example, to a colony of head-lice their whole world consists of the head of some unlucky individual.

Here their environment allows them to live, feed and breed. They don't need very much else to survive, and when the sufferer of this infestation, driven to desperation, applies some sort of de-lousing agent as treatment, it would cause their mass extinction!

During the course of its long history, planet Earth has had five major mass extinction events, that virtually wiped out all life on the planet. Some of these events go back hundreds of millions of years, but the evidence has been wiped out by erosion and the tectonic movements of the Earth's crust. At one time in Earth's early history, all land masses were one massive continent called Pangea, till movement of the tectonic plates broke this land mass apart and the continents drifted into the positions where they are today. It is not exactly clear what caused these mass extinctions.

At least one is thought to have been caused by an asteroid or comet impact, that was responsible for the extinction of the dinosaurs that dominated the planet for about 200 million years, and who disappeared all of a sudden 65 million years ago at the end of the Cretaceous period. Other explanations are changes in the Earth's climate, massive volcanic eruptions that caused great cracks in the Earth's crust, releasing molten magma and noxious gases, that not only killed off most animal life but most plant life too.

Although it may seem that our Goldilocks planet has evolved to make life comfortable for us, it is we (and that is life in general) that have had to adapt to the changes in the environment and take advantage of any that are favourable to us.

There is a hypothesis called the Gala theory, (conceived by James Lovelock) that suggests that organisms interact with their inorganic surroundings to form a complex and self-regulating system, and contributes to maintaining conditions for life on this planet. In other words, we have an influence in making the world a habitable place for us. At this point, perhaps I should take human beings out of the equation, as we seem to work against nature all the time by pollution which is gradually but slowly poisoning our atmosphere and oceans. Not only will we bear the consequences for this, but so will all forms of plant, aquatic and terrestrial life. According to Biblical terms, we are supposed to be the "guardians" of the planet, and we are not doing a very good job. We must work with nature if we wish nature to work for us.

The greatest mystery of all is how life came to be on this planet. Did it originate on this planet, or did it come from outer space? It can only be one or the other. In the first hypothesis, life could have evolved in our oceans, tide pools, hot springs, or hydrothermal vents deep on the ocean floor. Scientists are turning more and more to the latter supposition, as they have discovered that some types of organisms called extremophiles can survive in very inhospitable conditions, from freezing to almost boiling, and without the need of sunlight or, in some cases, even oxygen.

These hydrothermal vents could have also fuelled many of the chemical reactions for life to evolve. In the case of life being seeded from outer space, some scientists claim that two amino acids, adenine and guanine, have been found on meteors that have landed in the Antarctic. There is some argument as to whether these scraps of the DNA molecule, actually came from outer space or that the meteor was contaminated somehow, perhaps by a careless sneeze from someone in the vicinity!

It is conceivable that life could evolve deep in space, on something like a comet, or if not life itself, at least the ingredients of life, like linked amino acids. There is in fact a theory (the Panspermia theory) that life, such as extremophiles, can survive the hazards of outer-space, like radiation and the extreme cold, become trapped in space debris, and fall on a hospitable planet, like ours, to seed life. Some organisms can remain dormant for years on end, and then become activated when conditions are suitable for them to evolve. The seeds of some plants have this ability and can travel our oceans for many years, then germinate when they have found a suitable habitat. The Panspermia theory would be a way where all life could be spread across the Universe.

No doubt one day scientists will eventually discover how life evolved on this planet or elsewhere in the Universe. At the moment there is no definitive answer and "the mystery of life" will have to remain a mystery. It is always challenging for the end product to determine how it came to be.

If a wine glass could talk and we were to ask how it came to be would it have all the answers? Would it know that it was once part of a sandy beach? That it was melted, shaped and cut by some unknown hand? Such is life and such are the mysteries of life!

[27]

In Conclusion

EVER SINCE OUR ANCESTORS LOOKED UP AT THE STARRY NIGHT SKY AND WONDERED WHAT IT WAS, IN THE LAST HUNDRED YEARS OR SO WE HAVE AT LAST BEGUN TO COMPREHEND HOW VERY, VERY LARGE OUR UNIVERSE REALLY IS AND TO GAIN SOME SORT OF UNDERSTANDING HOW IT ALL BEGAN.

There is still a lot we do not know or cannot grasp, and no doubt new theories will evolve with more and more knowledge, observations and studies of this very massive subject. There are still so many unanswered questions. For a start, why is the Universe so BIG? What are we doing here? What is the purpose of it all? Why did intelligent beings like us evolve? How could a simple element like hydrogen evolve into the human brain? Not to be conceited, the human brain is probably the most complicated thing the Universe has ever produced – after all, as I have said before, most of it (the Universe) is just simple hydrogen gas. If you don't believe in miracles perhaps you will have to think again, because we are living proof of a miracle. It happened, and we are here.

But could we be the only intelligent species to marvel at the wonders of the Universe? Are we the only audience of a Shakespeare play, or is the auditorium empty with no one else to applaud? It seems inconceivable that we are the only ones, given the vast number of stars, and probably planets, in not only our own galaxy but in the billions and billions of others that exist beyond our own. Perhaps there are other civilisations not unlike ours, not simply living on a single planet but colonies of planets in perhaps the stars in star clusters where the stars are nearer to each other. But again we are faced with the awkward question that these would be so far away as not to matter.

The laws of physics seem to be the "Bible of the Universe." As far as we are able to determine, everything in the Universe appears to follow the laws of physics. These are very restrictive laws, which is why we can't suddenly flap our arms and begin to fly. Our imagination is in constant battle with the laws of physics, and we literally allow our imaginations to run away with us and create a world of fantasy where miraculous things can magically happen.

From the earliest written literature, like The Odyssey, supernatural heroes and mythical beasts have captured our imagination, and it is only when we are confronted with physical laws, that we come down to Earth with a bump. However, sub-atomic particles, according to quantum theory, don't appear to follow the laws of physics, which is baffling physicists into the realms of "fantasy" – like could a person exist in more than one place at the same time in parallel universes?

I hope that the foregoing chapters will not have boggled the minds of prospective readers, but instead given him or her a better understanding of the Universe, and what makes it tick. If this is the case, I would have achieved my objective!

About the Author

BORN IN 1939, JAMES SINCLAIR SPENT HIS CHILDHOOD AMONG THE SNOWY PEAKS AND CLEAR STARRY SKIES OF THE TOWN OF DARJEELING, IN THE HIMALAYAS DURING THE LAST DECADE OF THE BRITISH RAJ IN INDIA.

He attended the renowned Mount Hermon School, where the celebrated playwright Sir Tom Stoppard also once studied. Later, he worked in Calcutta in a commercial company, before migrating to the UK in 1969.

He worked in the PYE group of industries in Cambridge, then the Civil Service in Chessington, Surrey. He had to retire early due to heart problems, and found that retirement gave him more time to pursue his many hobbies, one of which was astronomy, and the other writing. He decided to put these two together and write Over Our Heads. However, like his many writing projects, this was put on the back burner for several years, and it was only by introduction to Tom Evans and his encouragement, helpful advice, and technical expertise that the book eventually took off the ground.

Always a keen fan of science fiction since a boy, James has grown up with the many sci-fi films from the early 1950s to the present day, and any science program screened on television.

He was a great fan of the late Sir Patrick Moore and seldom missed a Sky at Night program. Most of his understanding of the Universe has come from reading books and watching science documentaries, and the Internet has brought so much more information to the door.

He lives out his retirement peacefully in his Surbiton maisonette with his two cats, but wishes he was in the country where he could enjoy the clear night skies without the light pollution of city lights, and pursue his stargazing.

Acknowledgements

NO BOOK IS A SOLO EFFORT.

My sincere thanks to Jane Wilson for the sterling job she has done in proof-reading and editing this book for me, from far off Wellington, New Zealand. She painstakingly went through my manuscript and corrected any grammatical errors, and provided me with helpful suggestions and advice.

My thanks too to Tom Evans for his helpful advice and technical assistance in getting this book off the ground and published for me. His suggestions and ideas gave me food for further thought in improving the content of this book. His technical expertise in book publication and marketing has been invaluable, and if it weren't for his assistance, this book would have been like a rocket ship that failed to launch!

Thanks to Nathan Harris-Heigho for giving a younger person's perspective to the book and to Michael O'Connor for providing his thoughts from the other end of the age spectrum. You have both given me the feedback that I needed that this book works for all ages.

More Resources

YOU CAN FIND OUT MORE ABOUT ME, MY THOUGHTS AND MY WRITING ON MY BLOG SITE. DO DROP BY AND LET ME KNOW IF YOU HAVE ANY QUESTIONS, INFORMATION OR THOUGHTS THAT CAN HELP US BETTER UNDERSTAND WHAT IS OVER OUR HEADS.

Visit www.overourheads.co.uk to see what I am up to next.

www.ingramcontent.com/pod-product-compliance
Lightning Source LLC
Chambersburg PA
CBHW051502170526
45166CB00001B/357
* 9 7 8 1 5 1 9 7 7 3 0 5 0 *